WATER

BOOKS BY MARQ DE VILLIERS

White Tribe Dreaming
APARTHEID'S BITTER ROOTS AS WITNESSED
BY AN EIGHTH-GENERATION AFRIKANER

Down the Volga in a Time of Troubles
A JOURNEY THROUGH
POST-PERESTROIKA RUSSIA

The Heartbreak Grape
A JOURNEY IN SEARCH OF THE
PERFECT PINOT NOIR

Blood Traitors
A TRUE SAGA OF THE AMERICAN
REVOLUTION
(*with Sheila Hirtle*)

Into Africa
A JOURNEY THROUGH THE
ANCIENT EMPIRES
(*with Sheila Hirtle*)

Water
THE FATE OF OUR MOST
PRECIOUS RESOURCE

WATER

THE FATE OF OUR MOST PRECIOUS RESOURCE

MARQ DE VILLIERS

A Mariner Book
HOUGHTON MIFFLIN COMPANY
BOSTON · NEW YORK

First Mariner Books edition 2001

Copyright © 2000 by Jacobus Communications Corporation
ALL RIGHTS RESERVED

First published in Canada in 1999 by Stoddart Publishing Co. Limited

For information about permission to reproduce selections
from this book, write to Permissions, Houghton Mifflin Company,
215 Park Avenue South, New York, New York 10003.

Visit our Web site: www.houghtonmifflinbooks.com.

Library of Congress Cataloging-in-Publication Data
De Villiers, Marq.
Water: the fate of our most precious resource / Marq de Villiers
p.   cm.
Originally published: Toronto : Stoddart, 1999.
Includes bibliographical references and index.
ISBN 0-618-03009-3
ISBN 0-618-12744-5 (pbk.)
1. Water-supply. 2. Water-supply — Political aspects. 3. Water — Pollution.
4. Water resources development — Political aspects. I. Title
TD345.D473 2000
333.91 — dc21   00-021224

Printed in the United States of America

Book design by Robert Overholtzer

QUM 10 9 8 7 6 5 4 3

To the memory of
JOHANNES JACOBUS DE VILLIERS,
who farmed in a hard land and
knew water's true value

I float, therefore I am.

> — the Reverend Mother Darwi Odrade

Why repeat mistakes when there are so many new ones to make?

> — Descartes

Frankly, most of the academic studies [on water] are irrelevant to practical decision-making. But that's okay. They're academicians. They have no responsibility to manage real resources. We have to deal with real things — real dams, real rivers, real demands, real crises.

> — Eugene Stakhiv, U.S. Army Corps of Engineers

We're all downstream.

> — Ecologists' motto, adopted by Margaret
>   and Jim Drescher, Windhorse Farm,
>   New Germany, Nova Scotia

We used to think that energy and water would be the critical issues for the next century. Now we think water will be the critical issue.

— Mostafa Tolba of Egypt, former head of
the United Nations Environment Program
and Grand Old Man of the environmental
movement

Water is like the blood in our veins.

— Levi Eshkol, Israeli prime minister, 1962

Millions have lived without love. No one has lived without water.

— Turkish businessman, 1998

Whiskey is for drinkin'; water is for fightin'.

— Mark Twain

ACKNOWLEDGMENTS

For being helpful — nay, essential — along the
way . . . Rosemary Shipton, Earl Green, Ken Hirtle,
Tom Gardner-Outlaw, Elisabeth Cherdel and Kay Rolland,
Bill Gilkerson, Dominique Henry, Tom Tait, Gregory Grammer,
Helen Marquard, Maurice Strong, Kim Peters, Stephanie Foster,
Fraser and Linda Farmer, Professor Ben Dziegielewski,
Professor Henry Kendall, and, of course,
Bruce Westwood.

Contents

Preface

Behind the barn on our farm near Maynooth, in the deciduous forest belt of Ontario, is a small spring. It bubbles sleepily from the ground, and if you're really quiet and there's no wind in the trees, you can hear it making little burping noises, like a baby content at the breast. The water seeps through the grass and trickles into a stone runnel, left there by a farmer long gone, and then forms a pool, ducks underground for a while, resurfaces in a small wetland, and disappears into a ravine. Lower down it becomes a creek and, joined by others, a stream, a lake, and then . . . well, you know the rest. It fetches up in the sea, where it lives awhile.

When we first bought the farm, I used to take a chair outside in the summer sun and watch the water move. My Canadian friends thought it a risible thing to do. Water was water, and it was everywhere, wasn't it? I paid no attention, but stared down at the little spring. It bubbled and seeped and gurgled, and it was cold when I reached down to touch it. I had never "owned" water before. I grew up in the arid center of the South African plains. It hardly rained there (though when it did, the clouds burst), and for most of the year the rivers were dry, dusty places where thorn bushes grew and weaverbirds made their nests. My grandfather had water, but his borehole went down into the center of the earth, drawn to the surface by a clanking windmill, a charmless mechanical thing. For a while I thought it made the water, somewhere down there, in its rusting heart. Now, here was a wonder — water bubbling into the clear air, all by itself.

Later my aunt came to visit. She was a de Villiers, like me, and a Boer to her marrow. I drove her up from Toronto, taking the back roads through the endless forest, and she spotted at once the fundamental difference. "Here," she said, "you know when you've come to a farm, for there's a clearing. At home you know when you've come to a farm, for there are trees."

One place has water in abundance, the other not. One place pays little attention to the marvel of it — it's just "there"; the other is by necessity obsessed with it.

I understood when I was just a child that without water, everything dies. I didn't understand until much later that no one "owns" water. It might rise on your property, but it just passes through. You can use it, and abuse it, but it is not yours to own. It is part of the global commons, not "property" but part of our life-support system.

In the thirty or so years I've been traveling about the globe for one journalistic purpose or another, I've been "collecting" water, whether as dams or rivers or lakes. If time and projects permitted it, I would always allow myself to be diverted by water. And so I followed the Niger through the Sahel to Timbuktu and Gao; I stood on the lip of Kariba and watched the hippos wallow; I puttered through the Volga's delta, where I saw the sturgeon dying; I watched the dolphins splashing in the delta of the Nile; I watched the water hyacinths drifting by on the once-mysterious but still potent Congo; I stood on the vertiginous lip of the Hoover Dam; I walked across the Loire on the very stones used by Richard the Lion-Hearted; I took the ferry across the St. Lawrence at Rivière-du-Loup, and then took it back . . . But I wasn't a water snob. Any little stream would do. The LaHave River in Nova Scotia pleased me just as much, or Hard Labor Creek in South Carolina (on whose muddy banks my wife's ancestors pioneered after a terrible journey from Germany), or Burleigh Falls in Ontario, where I watched the fly-fishermen balancing on the rocks in the rapids, but never saw them catch anything. I would stop for mighty rivers and small creeks, for Great Lakes or muddy little puddles.

So I thought when I started this book that I understood water. But I didn't — not really. I knew where the water for my small spring came from, and how the forest cover sustained the water table that made it possible, but I had no real grasp of the intricate interconnectedness of the global hydrological system. I knew that the trees on my farm protected the groundwater, but I never knew the scale of the clearing of forests and wetlands that was occurring around the globe, and how that clearing led to alternating bouts of drought and flood. I could see for myself how industrial effluents were pouring into, say, Russia's Volga, but I didn't understand the wholesale assault our species is making on the system that sustains us. I always knew how scarce water was in many parts of the world, but I didn't really comprehend the calamity facing many regions of our increasingly overpopulated planet.

I hope I understand it better now. As you will see in the text, many wise people have been there before me and have aided my search; I have tried to indicate how valuable I found their company. Academicians and scientists all over the world are studying water, its availability and quality, its uses and abuses. Published sources range from the intricately technical (the permeability of schist formations) to the ecological and political (transborder water and resource conflicts). There are hundreds of books on the subject, and quite literally thousands of papers. But in the flood some names stand out, a helping hand for those of us floating by.

Among them is Sandra Postel, the nearest thing in the water world to a philosopher — always thoughtful, cogent, intelligent, with a point of view as limpid as a forest pool. Her *Last Oasis* is the call to arms most often cited by others in the field. Peter Gleick is the organization man of the water world: his books and papers are meticulous and precise, a fund of data for those who need it. My views on Russia's Igor Shiklomanov are expressed near the end of my final chapter; I came to see him, in the years I spent in the water world, as a central figure, authoritative and magisterial. Leif Ohlsson's book *Hydropolitics*, which covers much the same ground as this volume though aimed at the development community, has drawn together a stellar list of academicians, and I have cited some of them in the text. Malin Falkenmark, a Swede, is the person most often quoted by other hydrologists; her work is at the core of modern water research. For cogent summaries of technical hydrology and the hydrosphere, the *Encyclopaedia Britannica,* in both its print and on-line forms, remains essential for the amateur. It is often neglected by academics, who sometimes forget the meticulousness of its science.

Of the journalistic accounts, Marc Reisner's classic *Cadillac Desert* stands out. It deals primarily with the American Southwest, but it is almost compulsively readable, and folded into the enjoyable polemic is a huge amount of information. There are many other passionate and readable books on water, among them Joyce Shira Starr's *Covenant over Middle Eastern Waters: Key to World Survival.* Lester Brown's WorldWatch Institute (from which Sandra Postel emerged) has produced other writers of note, among them Janet Abramovitz and, of course, Brown himself, always a thorn in the side of the complacent. Henry Kendall, the Nobel laureate whose own main focus is food, gave me endlessly good advice. Population Action International, an NGO based in Washington, D.C., while not primarily focused on water, has produced a good deal of meticulous research. On the skeptic's side, I was drawn to Eugene Stakhiv of the U.S. Army Corps of Engineers, an engineer who writes (and talks) in clear, jargon-free, persuasive prose.

I begin the book in a country very remote from many of the people who will read it. Sophisticated travelers will know, of course, that the Okavango Swamp, in the dusty African country of Botswana, is one of Old Africa's last great places. But I started there not just for that reason. For me the Okavango illustrates more than almost any other place the ways in which water has become imperiled, not through the deliberate actions of evil men, the corporate rapists of ecological fantasy, but through the small doings of many — far too many — ordinary people, doing things the way they have always done them. That's where the real danger lies.

PART I

The Where, What, and How Much of the Water World

1

Water in Peril

*Is the crisis looming, or has it
already loomed?*

THE LITTLE *MOKORO*, a boat roughly hewn from a *mopane* log, drifted slowly through the waters of the Okavango Delta. It was tiny — hardly larger than the crocodiles whose snorkel-eyes could sometimes be seen, mercilessly obsidian. The water was a startling blue with eddies of silt, drifts of ochre and dun, easy enough to examine, for the boat rode only a few centimeters above the surface. There were sudden splashings from a nearby papyrus island as hippos rolled in the muck.

You could spend days poling through the Okavango's twisty hippo-ways in a *mokoro* and never see anyone. You can clamber out of your *mokoro* onto an "island" of swamp grass a meter thick, but if you were accidentally to plunge through — an easy thing to do — there would be a couple of meters of water beneath you. Everywhere improbable vegetation grows prolifically. There are papyrus beds, swamp grasses sharp as razors, and exposed roots worn smooth by heavy bodies passing.

The Bayei people, swamp dwellers, find their way effortlessly through this maze of identical channels, but sometimes even they go into the swamp and never return. A legend says there are legions of screaming skulls in the muck beneath the islands; at the end of the world, when the waters dry up, they will be exposed to confirm the apocalypse so long forecast. At twilight, when the hippos return to the channels, their heavy, lethal bodies cutting Vs into the water, it's an easy legend to believe.[1]

Now the Bayei are being told that the end of their world may come earlier — not through catastrophe as prophesied, but through the thirst of that most voracious and expansionist of species, humankind. The pipelines, dredgers, and cadres of water management engineers and the bureaucrats of the water commission draw ever closer, persistent as a virus with no cure, ready to suck the lifeblood from the delta. And in 1996, for the first time in many years, the annual Okavango flood never reached the delta. The rains had failed in Angola, the Bayei were told.

"How deep is it here?" I asked Kehemetswe, the Bayei who was poling the *mokoro.*

"Two meters," he said. "That's fifteen feet," he added helpfully, getting it wrong.

He wore baggy khaki shorts, a sun visor marked "All England Tennis Club," and 10-centimeter earrings made of chipped bone and braided twine. His pole was a skinny thing that bent perilously in the mud as he pushed us along. He said he would use it to thwack crocodiles if they became a nuisance. *Thwack* — it had a reassuring sound.

"Two meters? Is that normal?"

"Last year, three. Year before that, three. Year before that, three. Three. Always three."

"What does this mean?" I asked, but I already knew the answer. And the next year I saw him quoted in a Johannesburg newspaper: "Namibia wants to build a pipe to take water from the river through the desert," he said. "If the water dries up, it will be the end of our lives. All the things of our lives solely depend on it."[2]

The Okavango, the third-largest river in southern Africa, rises in the moist tropical hills of Angola, where it is known as the Cubango, and flows for about 1,400 kilometers through Namibia and into Botswana; there it soaks into the flat plains of the Kalahari and spreads out in a dazzling array of channels that make up Africa's largest oasis and the world's most spectacular inland delta.

Once the Okavango Delta was a lake, but a geologic era ago some slight tectonic shift in the earth's crust drained the water into secret crevices. The Okavango River continued to pour down across what is now the Caprivi Strip from the moist hills of Angola's Benguela Plateau as it always had, but the lake had disappeared. The river became tangled in thickets of reeds, giant papyrus, and mud, and then just vanished. *Riviersonderend,* the Afrikaners called it, River Without End. There are many such rivers in the wild and desolate north, the Great Thirstland, but the Okavango was the grandest, most fertile, and most beautiful of all.

In good years the waters still overflow the marshes into the Boteti River and reach the parched, arid surface of the Makgadigadi Pans to the southeast. In the best years of all they seep northeastward into the even more remote and mysterious Linyanti lagoons. This marsh is one of the grandest places on earth. So many legends, so entangled with mundane and exotic fact! The *munu,* the black tree baboons that folklore says stalk Okavango women, longing for the day they might be men again. Lions hunting at night. Lagoons boiling with hippos. Endless forests of thorn acacia and *mopane* trees, home to leopards, cheetahs, wild dogs, Cape buffalo, giraffes, a shopping cart full of monkey species, the greatest migrations of zebras on earth, and huge herds of elephants. These 60,000 elephants are the subject of bitter controversy. The Botswanan government prohibited all elephant hunting in 1983, to universal acclaim, but since then the herds have proliferated, causing massive habitat destruction. And they are now being culled: some 15,000 elephants will have to be shot, here and in Zimbabwe and Angola, to end their own suicidal eating binge.

The Okavango Swamp has other enemies: drought, for one. In the 1990s the Kalahari region and Angola suffered from the worst drought of the century, and no one knows whether the marsh will recover. El Niño was blamed, but when the disturbance ended and the rains still did not come, human-induced climate change became the cause of choice. Mostly, the inhabitants were reduced to praying for rain, but the water levels continued to shrink. Now from the south, from Botswana's orderly and well-managed cities, comes an enemy even worse than drought: growing human populations and their demands for more, more, more. And from neighboring Namibia, too. Namibia is a desert country and always parched. But its population is growing, and in 1996, with the country's reservoirs standing at 9 percent of capacity, Namibian authorities turned their eyes eastward, to Botswana's Okavango Delta.

By 1997 there were already threats of a "water war" between Namibia, which wanted to take 20 million cubic meters a year from the Okavango system, and Botswana, whose own dams and reservoirs were critically low after ten years without rain. In all likelihood this talk of war was mere hyperbole. Still, there were threats of sabotage if a pipeline proposed by Namibia were ever to be built. Late in 1997 residents of Maun, in Botswana, walked out of a conciliatory meeting called by the Namibian government, and both countries went to the International Court of Justice at The Hague to dispute a minor boundary issue, a sure sign of internation fractiousness.

In 1996, when the annual floods failed, Maun was put immediately at risk — not just tourism but the villagers, too, who need the river for washing, for fish, and for the water lily roots and reeds they use to build houses. The town's drinking water, drawn from boreholes, was also drying up. Botswana's hydrologists scrambled to find out just how rapidly the water table was dropping. Alarmist stories were heard everywhere. Namibia's pipeline would mean permanent drought, residents were told. Wells and boreholes would go dry.

Namibia, in turn, had its anxieties. Few places on earth are drier than Namibia. Of the meager rain that does fall, four-fifths evaporates immediately. Only 1 percent recharges groundwater tables.

Worse, Namibia has no perennial rivers, only seasonally flowing ones that are reduced to a trickle for several months and dry up completely in others.

To augment the water supply, Namibia has been tinkering with other options, including desalination and pumping groundwater from its fossil aquifers. But desalination, expensive enough anywhere, is prohibitive in Namibia, where most of the population centers are inland. Overpumping groundwater has also caused dangerous increases in water and soil salinity, as well as the rapid depletion of the aquifers themselves.

This chronic water shortage prodded Namibia to launch a planning process to extend its already massive network of supply pipelines and aqueducts, the Eastern National Water Carrier, to the Okavango River, which runs throughout the year along its northeastern border with Angola. The first phases of the plan would divert an estimated 20 million cubic meters of water annually (700 liters a second) from a point on the Okavango River near Rundu in northern Namibia — well before the river gets to Botswana — and pump it uphill through a 250-kilometer pipeline to Windhoek, the capital. Namibia sees the pipeline as the only feasible solution to keep pace with the water demands of its growing urban centers. Clearly, the need for both governments to negotiate a long-term solution is urgent.[3]

Not surprisingly, the Namibians have defended their plans and maintain that the amount they will be taking is negligible, even though they intend to quadruple or quintuple it to somewhere around 100 million cubic meters a year. Even that, they maintain, would draw off only 1 percent of the water flowing through the Okavango system. Richard Fry, Namibia's deputy secretary of water affairs, was blunt. "The severity of Namibia's water crisis leaves us with little option," he said after the Okavango floods failed in 1996. The dams supplying Windhoek and the central areas of Namibia were at an all-time low, and were expected to run completely dry in a year or two. Namibia's senior water engineer, a bluff Afrikaner called Piet Heyns, was blunter still: "If we don't build the pipeline and the rains fail again . . . we'll be in the shit," he told the press.[4] The corpses of 60,000 head of cattle, dead of thirst, littered the landscape.

What, indeed, are the alternatives? It's not that Namibia isn't trying. If desalination is too expensive, and pumping water from disused mines is only a temporary solution, what is left? Namibians are not profligate users of water. In the last few years, Windhoek's residents have been cutting back water use, achieving a 30 percent saving. The city's annual consumption of 17 million cubic meters has not increased significantly since independence in 1990, despite a growth in population from 130,000 to 220,000, and is now only one-third of the water consumed per capita in another desert city, Las Vegas. Significant increases in the efficiency of use — what Sandra Postel has called poetically the "last oasis" — are therefore unlikely, and, despite conservation, the country's water needs are expected to double by 2020. Where is this water to come from? Namibia has agreed to an extensive environmental impact study before spending more than half a billion dollars to dip its pipeline into the Okavango. But what happens if the study says the delta would be irreparably harmed? What happens if Botswana furiously objects? Richard Fry believes that Botswana will see Namibia's water crisis in the light of "humanitarian need" and will ultimately respond sympathetically to the pipeline project. But if it doesn't, what are the Namibians to do? And if the Namibians go ahead despite objections, what are the Botswanans to do?

The nagging questions remain: Is it safe to interfere with the delta's only supply of water? The Okavango is robust enough to survive anything except the water being turned off. If that happens, a Garden of Eden would return to Kalahari dust, the wildlife would migrate or die, and 100,000 humans would be reduced to slum dwellers in cities already unable to cope. If it isn't safe to divert the water, is it necessary? Where, in the balance of competing interests, does natural justice lie? At what point does man's stewardship of the planet and its resources collide with man's own needs? What is the ethical position? Has it come to this: a stark choice between human misery and the destruction of one of the planet's most magnificent jewels?

And perhaps the most difficult question of all: In any ecology, beyond a critical point within a finite space, freedom diminishes as numbers increase. This is as true of humans in the finite space of a planetary ecosystem as it is of the Okavango elephants or of gas molecules in a sealed flask. The human question is not how many can possibly survive within the system, but what kind of existence is possible for those who do survive.[5]

Why should the world care what happens in this obscure debate between two minor-league African nations? Both countries are interesting enough, but hardly worth the world's concentrated attention. Namibia is, after all, the most arid country in the southern end of an arid continent. And Botswana? It is also a curiosity: a democratically run, sensibly governed, economically sound country that has eschewed grandiose development projects in favor of small-scale enterprise, schooling, and decent housing. But this is a country that has only one and a half rivers. It too is mostly desert. Of course these two places will squabble over water. What has that to do with the water-rich North?

As we shall see, the Okavango is the world in miniature. All the great themes that are being played out on the global scale with water — diminishing aquifers, dropping water tables, alarm about sustainability — all the issues that are facing more populous places much more critical to global peace, are here being traced in sinewy outline. Here is a rapidly growing population placing a strain on a fragile and finite resource — just as it is in North Africa, China, many parts of Asia, sections of Europe, and the southwestern United States. Here are humankind's competing imperatives, for food and for "development." Here is a simple example of the transboundary, supranational nature of water basin and water resource debates — just as is happening along the Nile, the Mekong, the Ganges, the Tigris and the Euphrates, the Jordan, the Rhine and the Danube, the Colorado, Rio Grande, and Columbia. Here is a small-scale example of growing interstate tensions over an increasingly anxious need for life's most critical resource. Here, poignantly, is the imminent and probably unpreventable destruction of a superb and precious ecosystem, with all its intended and unintended consequences — much like the Three Gorges on China's Yangtze River. Here are human politics brutally undisguised, and a sign that necessity will always trump ethics — as the Americans have done to the Mexicans over the Colorado. Here engineers are trying to solve problems not of their own causing, with predictably dismal results — as in China, Libya, or the planet's most expensive welfare system, California's waterworks. Yet here also, by way of contrast, is the possibility for international cooperation: water wars are not inevitable. Namibia, Botswana, and Angola have set up a commission to discuss water rights, just as the Indians and Bangladeshis have done in their squabble over the Ganges.

I was in the Kenyan town of Narok the night a group of Maasai *morans*, warriors going through their rites of passage to tribal elder, clashed with the thuggish national police of Daniel Arap Moi. The cause of the ferocious riot that followed is of no consequence — warrior exuberance had gotten out of hand, and the police had overreacted — but, to be safe, I left them to rattle their sticks and truncheons at one another and took refuge in a nearby village. There I was invited in by a family of Gabbra, who lived in a tiny four-hut complex 3 kilometers from the nearest well.

The senior woman of the household, Manya, invited me to stay and pressed on me unwanted gifts of food she couldn't afford. It was, I knew, a typical African welcome.

In return, I picked up one of the four yellow plastic drums piled in front of the hut — it had started its life as a bulk container for vegetable oil in some far-off industrial city — and offered to help her fetch water. The family laughed, politely, but it was obvious what they were really thinking: white people, *mzungu*, are so inept. Fetching water was woman's work. So I stayed behind with the men, who were smoking on a wooden bench outside one of the huts.

Later that evening Manya and her daughters returned, each with a 15-liter pail balanced on her head. They swayed down the trail, singing one of their working songs to pass the hours, as they had done that morning, and as they would do on the days that followed, and as they expected to do, if they thought about it at all, forever.

A little later we ate corn mash and fried banana and sucked on mangoes. I declined the water, partly out of politeness and partly out of fear. The well was an old one and had originally been used by fifty families. Now, four times as many drew water from it, and they had to go down farther every year. A few months earlier, Manya said, two men had descended into the well and had passed up the muck in buckets, deepening the well by the height of a man. The water was muddy and smelled unclean.

All over East Africa — indeed, all over Africa — it is normal for people to walk a kilometer or two or six for water. In the more arid areas, people walk even greater distances, and sometimes all they find at the end is a pond slimy with overuse. More than 90 percent of Africans still dig for their water, and waterborne diseases such as typhoid, dysentery, bilharzia, and cholera are common. The bodies of many Africans are a stew of parasites. In some areas the wells are so far below the earth's surface that chains of people are required to pass up the water.

In Mali a few months later I stayed in a village in which an American non-government organization (NGO) had installed a solar-energy pump and a galvanized storage tank; it was still working perfectly five years after it was put in place. In Niger, across the border, a similar pump had broken, and one night a child had opened the stopcock on the tank and the water had all run out, soaking into the parched earth. The child was beaten, but it was too late. The water was gone and the villagers all moved. They never returned.

A year after that I visited a family of walnut growers in California's Central Valley. They had a drilled well out back, but had recently had to refurbish it because the water had run dry. They were down to 230 meters before they struck water again. They didn't mind. Their trees and gardens were irrigated by water brought in from the California Aqueduct and supplied to them at 10 cents a cubic meter, far below the cost of either collecting or transporting it. They were careful water managers, however, and meticulously metered the amount of water given to each tree — not like their neighbors, who let the water flow freely in furrows and frequently forgot to close the sluices. Yet they were members of the local golf club, whose fairways and greens were watered and fertilized all summer to preserve the lushness that golfers demand. They saw nothing incongruous in their behavior.

The rainfall in that part of the Central Valley was only 15 to 18 centimeters a year, the same as on the Kenyan plains. The water table was even lower. No one in Kenya could afford to install the kind of pumps that would deplete a subterranean aquifer, or "draw it down" at unsustainable rates, faster than it could be naturally replenished. Same climate, same rainfall, similar families. But when Bonnie Schuch, the walnut grower, wanted to fill her swimming pool, she turned the faucet without a second thought. Manya had never seen a swimming pool. Lazing in the water had never figured in her dreams.

The trouble with water — and there *is* trouble with water — is that they're not making any more of it. They're not making any less, mind, but no more either. There is the same amount of water on the planet now as there was in prehistoric times. People, however, they're making more of — many more, far more than is ecologically sensible — and all those people are utterly dependent on water for their lives (humans consist mostly of water), for their livelihoods, their food, and, increasingly, their industry. Humans can live for a month without food but will die in less than a week without water. Humans consume water, discard it, poison it, waste it, and restlessly change the hydrological cycles, indifferent to the consequences: too many people, too little water, water in the wrong places and in the wrong amounts. The human population is burgeoning, but water demand is increasing twice as fast.

The environmental movement, accustomed by now to fits of gloomy Malthusian soothsaying, has forecast increasingly common collisions between demand and supply. Even officials of so sober an institution as the World Bank have joined the chorus. Ismail Serageldin, the bank's vice president for environmental affairs and chairman of the World Water Commission, stated bluntly that "the wars of the twenty-first century will be fought over water." Although he was roundly criticized for this opinion, he refused to disavow it, and has frequently asserted that water is the most critical issue facing human development. Former UN secretary general Boutros Boutros Ghali said something similar about water wars. So did Jordan's late king Hussein, who had obvious cause to mean it. Egypt has more than once threatened to go to war over diversions of the Nile. Water is in crisis in China, in Southeast Asia, in southwest America, in North Africa — indeed, in much of Africa except the Congo, Niger, and Zambezi basins. Even in Europe there are shortages. Drought is no longer a word alien to England, where water tables began dropping in the early 1990s. In many parts of Europe, downstream towns and cities are feeling the consequences of the careless alteration of age-old hydrological ecosystems, as rivers suddenly rage out of control, wetlands dry up, and contaminants enter the groundwater. Yes, even in Europe there is a crisis in water supply and management, as groundwater tables sink and rivers are reduced to a trickle or increased to a destructive flood.

Of course, there are skeptics, just as there are those who don't believe in the widespread notion that one generation is simply the earth's steward, holding it in trust for generations to come. These skeptics believe that the problem is overblown, and even if it isn't, it will surely be solved through human ingenuity and technological advances in the future.

But these people are a constantly shrinking minority. Everywhere you look, there are signs that the water supply is in peril.

- The level of the Dead Sea has plummeted more than 10 meters over the last hundred years. The river Jordan has been reduced to little more than a drainage ditch. In northern Israel, the Sea of Galilee, which gives much of the south its water, is shrinking and threatening to turn saline. In Gaza, overpumping is reducing the hydrological pressure, which is letting the seawater in, and the wells are producing water that is less and less potable. Already Jordan, Israel, the West Bank, Gaza, Cyprus, Malta, and the Arabian Peninsula are at the point where all surface and ground freshwater resources are fully used. Morocco, Algeria, Tunisia, and Egypt will be in the same position within a decade.
- About 250 million people inhabited the earth 2,000 years ago. By 2020 there will be 400 million along the North African shores and in the Middle East alone. And the water supply is shrinking as fossil aquifers are used up.
- The Sahara is expanding. A trivial few thousand years ago, hippos played where there is now only stone and scrub.
- Lake Chad — once, it was supposed, one of the sources of the Nile — is shrinking at a rate of nearly 100 meters a year. Already, in dry years, humans can wade across it safely, if they are wary of crocodiles and hippos.
- Water supplies in the Nile Valley itself — the cradle of civilization — are in peril. Egypt is an efficient user of water, but Egyptians are consuming virtually all the available supply, and the population is growing at more than 3 percent per year. There are a million new Egyptians every nine months.
- In millions of hectares of northern China, the water table is dropping at a rate of 1 meter a year. Irrigation — and its wasteful runoff — is blamed. Beijing can now supply itself only by diverting water from farmers, who give up farming and retreat to the cities — adding to the water demand there. Huge diversion schemes are afoot to bring in water from the water-rich and flood-prone south,

but this may not be enough, or may not be in time to match need to supply.

• In the Punjab and in Bangladesh, where there is flooding almost every year, the rate of drop in the water table is even faster than in China. Too many people, too little retained water.

• The water level in the once pristine Lake Baikal, the deepest freshwater lake in the world, is sinking steadily. At the same time, the quality of its water is deteriorating as effluent from unregulated factories pours into it.

• In Europe, although there are successes, most of the major rivers carry industrial and human wastes to the sea. Even in remote parts of continental Europe, the water from streams can be unfit to drink. In many parts of Slovakia, Poland, and western Russia, the rivers run yellow with industrial poisons.

• "There is far less good water to drink in England and Wales than was previously supposed, the Environment Agency claims in a report issued today." In these words the British media, in March 1998, reported that a supposed buffer supply of about 1 billion liters of water did not in fact exist. The study, which looked closely at rainfall patterns over recent years, found the northwest, Thames Valley, and west of England and Wales to be the hardest-hit areas, though there has been an increasing incidence of drought in general, and certain critical aquifers are shrinking. In the fourth year of the major drought of the early 1990s, *The Economist* reported, the margin between supply and demand in England had shrunk alarmingly to 3 or 4 percent. England — England! — was facing droughts.

• In the southwestern United States, politicians have notched the rhetoric up and are beginning to view northerners' reluctance to divert water southward as an act of ecological aggression — not just from northern California, Oregon, and Washington, but from Alaska and Canada, too. Some of the grandest rivers of northwestern Canada, in this view, are being "wasted" — allowed to flow uninterrupted into the oceans instead of being channeled southward to irrigate parched farmland. Las Vegas is demanding a greater share of the waters of the Colorado River. Many places in the High Plains are overdrafting the aquifers on which the region's farmers depend.

• Entrepreneurs in Colorado and other states have run into furious and passionate opposition to their plans to "mine" water; the private control of water resources is more and more an issue. In 1974, Roman Polanski's movie *Chinatown* had as its underlying theme the willingness of politicians and developers to murder for the right to bring water to the American Southwest — so valuable a resource did it appear. Since then, there have been several celebrated real-life civil trials involving the crucial question: Who controls supply?

It's now, of course, just a small curiosity of American history, but the last time one state took up arms against another was over water. In 1944 Governor Benjamin Moeur, a politician with a flair for the dramatic, dispatched the Arizona National Guard to the Colorado River during the construction of the Parker Dam, his declared intention being to stop California's "theft" of Arizona's water. A one hundred–man unit with machine guns mounted on trucks appeared at the dam site, and construction stopped, as well it might. The U.S. Supreme Court subsequently ruled in favor of Moeur's claim that California had been acting illegally in expropriating water without Arizona's permission. But Congress set the decision aside by passing a retroactive bill to legalize the theft, and that was the end of that.

At a 1998 UNESCO conference, "Water: The Looming Crisis," Peter Gleick, a short, neatly bearded, precise fellow who exudes a justified air of authority and competence, was up on the platform arguing passionately for different ways of looking at water. The crude global or continental measures are all very well, he was saying, but they tell us nothing about the human costs. Even national figures can be misleading, as we know. Parts of Bangladesh may be flooded, others parched. Or a country can be underwater one month and stricken with water shortages a few months later. Water availability is one measure, certainly. Actual water use — withdrawals from the system — is another, and it tells us different things, some of them difficult to interpret. A further way of looking at water is to see how we can provide for the basic water needs of all citizens. "The actual amount you give to each individual — the number of liters — is not that important," he said, "only that we move from *zero* to *something*."

Gleick paused and leaned forward on the lectern. "You ask me about costs?" he demanded. "I can tell you the cost of not providing basic water for drinking and sanitation will far outweigh the cost of doing so." These costs, he said, were currently running between $100 and $200 billion a year for health care and social welfare alone. "About half the modern world doesn't have the same basic amenities the ancient Romans took for granted."

Gleick recommended that UNESCO adopt a "human entitlement" of 50 liters of water per person per day. "Drinking water, 5 liters; sanitation water, 20 liters; bathing water, 15 liters; food preparation, 10 liters. Total, 50 liters." These figures, he pointed out, are far below even the minimal average withdrawals per capita in the most water poor of countries. "This is not a technological issue. The technology is easily available. It is a political and organizational issue. Water is a social good — we all agree on that. People should pay for its use, to encourage efficiency and as a recognition of its value. But perhaps a universal 'lifeline rate' should be established, and anything above that should be priced much higher. To water a lawn, for example, should be truly expensive."

I was only half listening. I was still puzzling over something else he had said about the amount of water withdrawals on a continentwide basis. Most hydrologists accepted Malin Falkenmark's notion that 1,700 cubic meters per person per year was the cutoff between a country being water stressed and reasonably comfortable. Yet when I looked at Gleick's figures for actual water use, as opposed to available water, no one was really using the full 1,700 cubic meters. Even the Americans and Canadians, the greatest wa-

ter hogs, were using only 1,693 cubic meters a year, out of a re-
source much greater than that. In descending levels of use,
Oceania (the Pacific Islands, Australia, New Zealand, Papua New
Guinea, and Fiji) used 907 cubic meters, Europe 726, Asia 526,
South America 376, and Africa a puny 244.

And I had just that morning finished reading a pamphlet put
out by Population Action International which had dealt with revised UN population estimates for 2050, projections made in 1996
that were sharply lower than those the United Nations had been
gloomily forecasting only two years earlier. Even with the downward revision of population forecasts, the "medium projection"
numbers showed that in a world of 9.4 billion people in 2050, a
billion people would be in "water scarcity" and 970 million in "water stress." Another report, published by Johns Hopkins School of
Public Health, had found that "caught between the growing demand for fresh water supplies on the one hand and limited and increasingly polluted supplies on the other, many countries are making difficult choices." How to square these circles?

When Gleick's lecture was over and he made his way to the coffee machine in the lobby, trailed by a wake of hydrologists from a
dozen countries, I stopped him and asked about the slipperiness of
the numbers. "Your book is called *Water in Crisis*," I said. "This
conference is called 'Water: The Looming Crisis.' How do you
match this 1,700 cubic meters a year with actual consumption and
come up with a crisis?"

"First of all," he said, "you have to be careful with these numbers. To some degree, water statistics are a technocratic illusion.
A thousand cubic meters a year doesn't necessarily mean water
stress. Israel is well below that, at about 300 cubic meters per person per year, and while they're on the edge, no one at present suf-
fers from a lack of basic amenities. In Nigeria, which has lots of
water, more than half the population goes without safe drinking
water." If a country draws less than its available resources, he suggested, it doesn't mean that it is living thriftily. It might simply
mean that the infrastructure is a shambles. He'd just been ap-

proached, he said, by *National Geographic* magazine, which was
planning a major feature on water and wanted to use some of
his data. "They made the same point to me. 'No one uses 1,700
cubic meters,' they said. But that's only one way of looking at it.
They were ignoring, for instance, the use of rain-fed agriculture.
When that's added in, many, many places approach 1,700 meters."
He shrugged. "You have to look at total resources, renewable resources, usable renewable resources, the ability to transfer water
from water-rich to water-poor places, the development level of
the economy, the annual consumption, and the deprivation level,
all matched against population trends and economic resources.
When you do that, you'll see that there are crises in many places.

"And another point. Forecasts should be scrutinized carefully
and used cautiously. In the 1960s, water consumption in America
was forecast to increase by up to 150 percent by the year 2000.
This has not happened for a number of reasons. A fundamental
change in attitude, for one thing. A shift from profligacy to conservation has stabilized demand, and withdrawals are in fact shrinking slightly."

Just before he turned to walk away, he repeated a point he'd
made on the podium. "It is difficult," he said, "to see the abyss before you fall into it. Are we really on the edge? Take Mexico City,
for example. They're providing basic water needs for their citizens.
But we already know that unsustainable pumping of the local
groundwater has caused parts of the city to subside by nearly 20
meters. They're now bringing water in from 300 kilometers away,
pumping it uphill a substantial distance. Population is still growing explosively. How close to the edge are they? How close to catastrophe? That's why we should be looking at these things. In many
places, unpleasant surprises are inevitable."

The human "need" for water depends on definitions. The crisis,
real though it be, is to some degree a management problem, a matter of allocation and distribution, and not just a pure problem of
supply, although in some places — North Africa, the Middle East

— it is that, too. Peter Gleick defines water needs as "access to basic drinking water and water for sanitation needs," which seems straightforward enough. By this test, and according to the latest data, most of Africa, most of Asia, and western South America fail. More starkly, over 1 billion people have no access to clean drinking water, and more than 2.9 billion have no access to sanitation services. The reality is that a child dies every eight seconds from drinking contaminated water, and the sanitation trend is getting sharply worse, mostly because of the worldwide drift of the rural peasantry to urban slums. Of course, Gleick's measure is personal and humanitarian, and doesn't factor in agriculture or industry. Other water experts define needs differently. Population Action International, for example, maintains that the number of people living in "water stress and water scarcity" was 436 million in 1997, and projects that the percentage of the world's population without enough water will increase fivefold by 2050. UN figures show it will be worse than that. Per Pinstrup Anersen, director of the International Food Policy Research Institute, says that one in every five countries is likely to experience a severe shortage within twenty-five years. Malin Falkenmark, whose figure for "water stress" of 1,700 cubic meters per person per year has been adopted by most hydrologists, nevertheless suggests that any nation with less than 1,000 cubic meters per person per year is "water scarce"; it takes 1,100 cubic meters to grow the food needed for one person's nutritious but low-meat diet for a year.

I reviewed my notes.

Europe has, on the whole, plenty of water, about 4,066 cubic meters per person per year, based on a 1998 population of 498 million. In only isolated instances are Europeans without access to safe water, and those are generally caused by civil war or by temporary pollution problems. But of course the average is skewed by the Scandinavian countries, which have water to spare (90,000-plus cubic meters per person in Norway, and a whopping 624,000 in Iceland). In the south, the situation is much more dire. Spain has only 2,800 cubic meters, much lower in the east and south, where consumption is passing critical levels. The south of France is not much better off, and the situation is compounded by the serious industrial pollution pouring down major rivers into the Mediterranean. The French, with two of the largest private water companies in the world, are beginning to charge realistic delivery costs to the affluent, with their swimming pools and Jacuzzis, in the hills of Provence.

North and Central America have, at first glance, water to spare. For a population of 427 million, there is an available and renewable water supply of 6.945 million cubic kilometers of water, or 16,260 cubic meters per person per year. But, again, the figures are crude. Canada has more water than the United States, by about half a million cubic kilometers, with a tenth of the population. Many parts of the United States have plenty of water — or would have if people weren't polluting so much of it. But in other parts they are draining aquifers by recklessly mining them dry, compounded by a snarl of laws and regulations designed for a simpler era, when natural resources seemed to be limitless. The United States has a theoretical availability of over 9,000 cubic meters per person per year, more than five times the stress level. Yet there are water shortages. Virtually all the available rivers have been dammed, and already more water is being shifted from one place to another than in any other country on earth, and major wetlands have been thoughtlessly drained. Still, there are positive signs: demand has been dropping — in certain places, such as Boston, substantially — and more thought is being given not only to the natural functions rivers perform but also to the restoration of wetlands, most notably in Florida.

Mexico is relatively parched, with a potential supply of a little less than 4,000 cubic meters per person. Parts of Mexico were always desert, but these areas are spreading throughout the northern part of the country because of misuse. Other human-caused deserts are extending in many parts of the Americas, including southwestern Utah and Oklahoma, parts of southern California, the southern half of Arizona, most of New Mexico, western Texas, and southern Nevada. The remaining soil in many of the same regions is rapidly becoming saline, impossible to cultivate even if the water were available.

South America averages a hefty 34,960 cubic meters per person per year for a population of 296 million, but the figures are hopelessly skewed by the Amazon Basin, the greatest reservoir and rain forest on earth and the greatest source of the planet's biodiversity. Paraguay is the only American country where less than 50 percent of the population has access to safe water. Peru has the least water in South America, with a mere 1,700 cubic meters per person of potential availability, and little Suriname the most. Suriname's admittedly tiny population of 420,000 people is awash in 468,000 cubic meters of water each.

Africa has a disturbingly low water resource potential of 6,460 cubic meters for each of its 650 million people, and even this paltry amount is inflated by the Congo River and the moist tropics. Africa also has the greatest desert on earth, the Sahara, which covers 8.6 million square kilometers.

Africa also has some of the greatest lakes in the world, among them Victoria, shared by Kenya, Tanzania, and Uganda, at 69,484 square kilometers; Tanganyika, shared by Burundi, Tanzania, Congo, and Zambia, at 32,893 square kilometers; and Nyasa, shared by Mozambique, Malawi, and Tanzania, at 29,600 square kilometers. Lake Chad, shared by Cameroon, Chad, Niger, and Nigeria, once measured more than 20,000 square kilometers, but by the early 1980s it had been reduced to 17,806 square kilometers and is still shrinking rapidly.

An average for Asia, with a population of around 3 billion and available water resources of 10,114 cubic kilometers, represents another crude measure. Some countries, such as Laos, have more than 55,300 cubic meters per person per year. But others, such as heavily industrialized Japan, are dependent on only 4,400 cubic meters per person. The critical countries are China, India, and Pakistan, which together account for more than 2 billion of the total population, and there the picture is much more dire: China has no more than 2,295 cubic meters per person, and that mostly in the south; India has even less, at 2,240, and is heavily dependent on the Ganges and the Indus rivers in the north.

Most hydrologists lump the Middle East (the Levant, the Arabian Peninsula, Iran and Iraq, and west to Turkey) in with Asia. But the politics of the region are unique, and so are its water problems, the most obvious of which are the fractiousness of Israel and its neighbors over shared water resources, and Turkey's role as upstream provider to Iraq and Syria. The region holds about 190 million people, 60 million in Turkey alone, and has a shared water availability of 370,000 cubic kilometers. Bahrain and Kuwait have no water of their own; at the other end of the scale, Iraq has 5,430 cubic meters per capita per year.

Australia, at 4.7 million square kilometers, is the sixth-largest country in the world. It is also the driest inhabited land mass on earth. It has the least river water, the lowest runoff, and the smallest area of permanent wetlands on the planet. Australia is not as barren as the Sahara, nor as arid, for even in the heart of it, in the region around Lake Eyre, there is still an average of 20 centimeters of rainfall a year. New Zealand has 91,800 cubic meters per person.[6]

But these figures don't necessarily tell us much about the world's flash points. How should they be calculated? There are a number of criteria: where the water supply is static or falling; where there is a dependency on water supplies from outside national boundaries; where rainfall is unsteady or meager; where populations are increasing; and where there are incompatible demands for water from competing internal sources (agriculture, basic population needs, industry). In Africa alone, by these measures, 300 million people, one-third of the continent's population, already live under conditions of scarcity, and this number will likely increase to more than a billion by 2025. Nine of the fourteen nations of the Middle East already face water-scarce conditions, and populations in six of them are projected to double in the first twenty-five years of the twenty-first century. India could join the list of water-scarce countries by 2025, almost entirely because of population increase. China, with 22 percent of the world's population and only 6 percent of its fresh water, is in serious trouble already: one-third of the wells in the northwest have gone dry, and more than three hundred towns have suffered water shortages.[7]

In 1992 Sandra Postel calculated that "if 40 percent of the water required to produce an acceptable diet for the 2.4 billion people expected to be added to the planet over the next thirty years has to come from irrigation, agricultural water supplies would have to expand by more than 1,750 cubic kilometers per year — equivalent to roughly 20 Nile rivers, or to 97 Colorado rivers. It is not at all clear where this water is to come from."[8]

Worldwide, more than three hundred river systems cross national boundaries. Hardly any of the world's major rivers are contained within the borders of only one state, and even fewer now that the world's last great empire, the Soviet Union, has broken up. Watersheds seldom acknowledge humankind's political conceits and pay little attention to frontiers. Downstream problems are not always solvable if upstream is in another country. Were Ethiopia to divert or use substantial portions of the Blue Nile, Egypt, entirely dependent on the Nile for its moisture, would be starved of water, and Egyptian politicians have always made it clear that they would have no option but war were that to happen.

Wars, or threats of wars, have been made in several riparian systems. The water resources of the Golan Heights and Gaza have figured largely in the military minds of Israel and its neighbors. The Jordan, Yarmuk, and Litani rivers have all been subject to military planners, and Israel has always treated water as a matter of national security. Water, and the Indus in particular, has poisoned relations between India and Pakistan. India and Bangladesh squabbled for decades over the Ganges, and though both these disputes have been tentatively resolved, there are several unsettling internal water issues that have frequently threatened to end in violence and have several times spilled over into riot, murder, and assassination:

militants from Tamil Nadu state have threatened guerrilla warfare on neighboring Karnataka, and Sikh separatists have manipulated water issues to their gain. Iraq, Syria, and Turkey have each mobilized troops in defense of water rights on the Euphrates and Tigris. In Europe, upstream "grooming" of the Rhine and the draining of its safety net, the Rhine wetlands, have caused downstream flooding; industrial pollution is another irritant. The United States has essentially "stolen" the Colorado from Mexico, using much of it to irrigate the deserts of Arizona and California, but a good deal of it to fill swimming pools in Los Angeles and fountains in Palm Springs. The Paraná, dammed and flooded, has caused friction between Argentina and Brazil.

Only one-third of the water that annually runs to the sea is accessible to humans. Of this, more than half is already being appropriated and used. This proportion might not seem so much, but demand will double in thirty years. And much of what is available is degraded by eroded silt, sewage, industrial pollution, chemicals, excess nutrients, and plagues of algae. Per capita availability of good, potable water is diminishing in all developed and developing countries. In the gloomy forecast of an eminent food bureaucrat, "Worldwide use [of water] has become so excessive that the implications for irrigated food production are considerable."[9] As Mostafa Tolba, former head of the United Nations Environment Program (UNEP), put it in 1998, "Just to match demand, major water projects will have to be started within the next ten years, or global supply will be overtaken." But most of the "easy" sources of water have already been exploited, and much of the water is in places where it isn't needed. Demand, it seems, will inevitably intersect with supply. And then what?

Will we find the resources through conservation and increases in efficiency — Sandra Postel's "last oasis"? Will we find it through heroic engineering (in bigger dams, longer pipelines, and greater desalination plants), or in the invention of new technologies such as fusion power?

Namibia and Botswana will probably not go to war over water.

What would be the use? Botswana will concede a little, Namibia reduce its demands a little, and the crisis will be staved off. Only the Okavango — an ecosystem in balance since man was a little hominid scavenging on the savannas of East Africa — will suffer and be diminished, a jewel that will glimmer less brightly.

2

The Natural Dispensation

*Who has how much, and who's
running out?*

THE ICEBERGS DRIFT PAST the harbor entrance at St. John's, Newfoundland, massive, beautiful, and deadly, but only to vessels blinded by hubris, incompetence, or sheer rotten luck. In the aftermath of the dizzying success of the movie *Titanic*, Concorde-loads of affluent Europeans jetted in to the Newfoundland capital to gawk. They would go out in Zodiacs, circling the bergs, some of them as large as small islands drifting in the Labrador Current, stately as dowagers. The Europeans would snap away on their Nikons, but the photographs, when they got home, never did them justice — they never showed the awesome scale, the pristine beauty, the blue glow in the sunshine. Even when the photographs showed the icebergs drifting past the harbor entrance, with human habitation in the background for comparison of scale, they never looked as they did at sea — pale, frigid sapphires, the breeze in the summer sun frosty from their passing.

Icebergs smaller than houses the Newfoundlanders contemptuously refer to as "bergy bits." One such, no bigger than a pickup truck, was seen dashing itself against the rocks in the fishing village of Upper Island Cove on Conception Bay. We'd been talking about icebergs that afternoon.

"You really should taste the water," said Ron Whynacht, who is from Lunenburg, Nova Scotia. "Twelve-thousand-year-old water. The cleanest, purest water you'll ever have. Drink it clean, by itself. Or twelve-thousand-year-old water mixed with twelve-year-old Scotch."

The floes that drift past Newfoundland have broken off from Greenland glaciers. They began there as snow in prehistoric times; the cold and the dry air inhibited evaporation, and there they stayed. Snow fell year after year, and, in the centuries and millennia that followed, the snow was compressed, incorporating some of the air from the original snow as bubbles, which disappear only through compaction at depths exceeding 1,000 meters. The Greenland glaciers, like most of the world's ice rivers, have been in a fairly steady state since the last ice age, gaining as much through new accumulation as they lose through breakaways. The ice moves slowly toward the ocean, not just because of basal sliding and the pressure of new ice forming but because of the peculiar nature of the ice itself, a crystalline solid which can "flow" just as water does, only more ponderously, as consistently as any river but infinitely more slowly. The glaciers can be hundreds of meters thick and might grind the bedrock as they move. When they retreated last time, they left behind moraines, the ground-up debris of their passing, and for centuries farmers have planted there, thinking little of how the soil came to be.

Glaciers may well retreat further if the dire warnings about planetary warming are true. Already certain glaciers, such as the Muldrow and Variegated glaciers in Alaska, sometimes surge rapidly and occasionally catastrophically, for no apparent or known reason. British Columbia's Athabasca Glacier has retreated 2 kilometers in a hundred years, and the rate is increasing. In Alaska's interior, the thawing of the permafrost has caused what the locals call "drunken forests," the trees tilting and leaning as the ground subsides. And more recently and alarmingly, some of the Antarctic's major ice shelves have suddenly shrunk. In March 1998, for example, the Larsen B ice shelf abruptly lost a 200-square-kilometer block that collapsed into the sea, and the following November an even larger section, nearly 1,800 square kilometers, disappeared, leading scientists at the U.S. National Snow and Ice Data Center to say, gloomily, that "this is the beginning of the end." This forecast might seem overwrought had not the Larsen A shelf, all 1,300 square kilometers of it, not disintegrated entirely in 1995, and the Wilkins shelf, more or less the size of Delaware, was expected to be gone entirely within a few years.[1] And at the turn of the century, the journal *Geophysical Research Letters* published a report by a group of oceanographers that found the Arctic ice melting far more rapidly than anticipated. The ice has thinned out by 40 percent in four decades — "A startling result," according to Andrew Rothrock, the scientist who led the study.

In Upper Island Cove, Ron was still talking about iceberg ice. "It will fizz and crack when it melts, or when you drop it into a drink," he said. "Crystals under pressure."

We were staying at a small inn on Conception Bay run by Barbara and John Mercer. John was a man who knew how to take a hint; he looked at the bottle of Scotch on the table and went to fetch a bucket and an ax. Ron and John scrambled down the 50-meter cliff and balanced precariously on the sea-washed rocks to grapple a bergy bit fragment, about twice the size of a human head. They heaved it into the yellow plastic pail and took it back to the inn. While they changed their sodden boots and water-soaked pants, another guest, Jim Lockhart, went at the block with a kitchen knife and a cleaver and split it into ice cubes.

Radioisotope dating had put the age of a similar chunk at somewhere between eight and ten thousand years old. I looked it up later. It had a pH balance of 5.4. I knew that pH, the negative logarithm of the hydrogen ion concentration in moles per liter, ranged from 0 to 14, with the lower numbers denoting increasing acidity, which made this ancient chunk slightly acidic but not as high as, say, acid rain. There were minute traces of ions of potassium, sodium, magnesium, calcium, chloride, sulfate, and bicarbonate. There was also a tiny amount of dissolved silica — about 0.30 parts per million. Hardly any water is truly pure. Only distilled water is just H_2O, and the hydrologic cycle is not as simple as I had first supposed. How did minerals, even in minute amounts, get picked up by evaporation?

Jim dropped chunks of ice into tumblers. They were clear — as clear as, well, spring water. The blue of the bergs was no illusion, but it was the refraction of light that drew the color of the sea into the crystals of ice. Jim added a sprinkling of white wine to one of the glasses. The ice crackled a little, but no more than ice made from tap water. It didn't mean that Ron had been wrong about the crackling. It more likely meant that the ice was older than he'd thought, and from deeper down, possibly deeper than 1,000 meters.

I put a small piece into my mouth and crunched. The water slowly melted. I tried a bit of tap water on the side. After the iceberg water, it tasted heavy, pregnant with additives. Chlorine traces, probably. Concerns for health will beat those for taste any time, and the public utility people have other things to worry about than pristine clarity.

In a good many parts of the world, the real worry is finding any water at all.

It's already a cliché in hydrological and Green circles that water is at once our most precious and most abundant asset. Many years ago Adam Smith pointed out that water, which is vital for life, costs nothing, whereas diamonds, useless for life, are valued highly. Water is everywhere. Humans, if rendered down, are 70 percent water. Frank Herbert, in his series of ecological novels about the desert planet Dune, was psychologically apt when he had the inhabitants, the Fremen, "taking back the tribe's water" in the rendering tanks when one of their number died. All life depends on water; indeed, life probably began in water. Water's curious heat-retaining properties steady the climate and make life on our planet sustainable. Without clean water, disease and misery take their toll. Without water we die.

It is also a cliché to say that the world is not short of water but rapidly running short of usable water.

Where did water come from? The most common assumption is that the earth itself is around 4.6 billion years old, formed by gravity from cosmic junk, clouds of ionized particles around the sun, debris left over from the somewhere-sometime explosion called the Big Bang. This cosmic garbage dump coalesced to form a protoplanet, which grew by gravitational attraction of even more junk (what the cosmologists call "particulates"). This was the Hadean Eon: a sort-of earth existed, but there was no atmosphere, no ozone layer, no continents, and no oceans — and most definitely no life. Around half a billion years later, give or take an eon or two, things had settled down enough to precipitate rocks; the oldest known rocks are in Greenland, and have just celebrated their 3.9 billionth birthday. The earth was still aflame with volcanoes and bombarded by asteroids, meteorites, and whatever else was floating in the interstellar void and intersecting with our nascent planet, but those oldest rocks show signs of having been deposited in an environment that already contained water. There is no direct evidence for water for the period between 4.6 and 3.8 billion years ago. But suddenly, there it is.[2]

The prevailing theory is that the atmosphere was created from the release of gases from volcanic eruptions. As the eons passed, the first lightweight silica and aluminum rocks, which are typical of continental land masses, formed. The surface of the earth cooled, and water vapor in this spanking new atmosphere condensed to form the water of the oceans.[3] Which begs the question: Where did the water vapor in the atmosphere come from in the first place? What was it in the volcanism that was our first "weather" that produced water? Or perhaps it was already present in all that cosmic junk — comets, after all, are sometimes little more than frozen lakes of water — and it came from space, an alien and infinitely curious little molecule. For water *is* curious, much more curious than it might at first sight appear, and is actually little understood. Why is it, for instance, that water is the only substance whose solid form is less dense than its liquid one (a phenomenon that has profound implications for aquatic life)? Scientists have discovered that water is made up of hexagons of hydrogen atoms arranged in what is called a "cage" structure, and that the smallest theoretical drop of water is made up of six molecules arranged as a cage. Which means what? No one really knows.

> As a liquid, water has special thermal features that minimize temperature fluctuations. First among these features is its high specific heat — that is, a relatively large amount of heat is required to raise the temperature of water. The quantity of heat required to convert water from a liquid to a gaseous state (latent heat of evaporation) or from a solid to a liquid state (latent heat of fusion) is also high. This capacity to absorb heat has several important consequences for the biosphere, including the ability of inland waters to moderate seasonal and daily temperature differences, both within aquatic ecosystems and, to a lesser extent, beyond them.[4]

In any case, water appeared, precipitated from what were not yet called the heavens. Then, around 2.5 billion years ago, life on earth began. And it began, almost certainly, in water.

Darwin and the early evolutionists imagined life evolving in a pool of soupy, chemical- and nutrient-rich water, an idea still pretty much accepted today, although there is a small but influential scientific subset that believes that life, too, might have come to us from space, ready formed, cosmic nuggets among the infinite dross. "Runoff collected in a small volume is the most likely means of achieving the necessary concentration of ingredients," says Gustaf Arrhenius, a geochemist at Scripps Institute of Oceanography. In the dry language of chemists he writes: "Ponds may have further concentrated compounds on the internal surfaces of sheet-like minerals, which attract certain molecules and act as a catalyst in the subsequent reactions. Two aldehyde phosphate molecules thus united form a sugar phosphate, a possible precursor to organic life" — as though that explained anything at all.[5]

How much water is there?

In recent years there has been increasing evidence that water is far from scarce in the universe. In March 1998, for example, a meteorite that fell to earth in Texas (breaking up a basketball game in the process) was found to contain water inside crystals of rock salt.[6] We now know that Europa, a pretty little moon of Jupiter, holds more water than all the oceans of the earth, a "shell" of liquid and frozen water 160 kilometers thick. "Water is very common in the outer solar system," says Ronald Greeley, a planetary geologist from Arizona State University.[7] Water traces have been found in the sun, the moon, the planets and their moons, asteroids, and even distant stars. "Water molecules must have been a major constituent of the solar nebula from which the planets formed," according to Robert Clayton, a physicist at the University of Chicago. A huge cloud of oxygen, hydrogen, and other gases in the constellation Orion pumps out enough water molecules to fill the earth's oceans sixty times a day.[8]

The total amount of water on our planet has almost certainly not changed since geological times: what we had then we have still. Water can be polluted, abused, and misused, but it is neither created nor destroyed; it only migrates. There is no evidence that water vapor escapes into space — a popular theory for those speculating about the fate of, say, Mars, but out of fashion since the moon, with no atmosphere of its own, was found to contain significant amounts of water in certain deep crevasses, "somewhere between 2.6 and 80 billion gallons" (a nice tidy 10 billion gallons translates into a shower for one person lasting about 3,800 years, or 12,500 Olympic-sized swimming pools, or one rather small lake — a tiny amount by earthly standards but ample for a small moon colony). And in late 1998 NASA announced a new mission — to "map" the water on Mars. A small amount of water, called juvenile water in hydrological circles, flows to the surface through "outgassing" of the mantle in "geologically active zones" (aka volcanoes), but it is thought to be balanced by water consumed in the hydration of minerals.[9]

The hydrologic cycle describes the way water circulates through the earth's systems, from a height of 15 kilometers above the ground to a depth of some 5 kilometers. It is a self-regulating quasi–steady-state chemical system that transfers water from one "reservoir" to another in complex cycles. These reservoirs include atmospheric moisture (cloud and rain), the oceans, rivers and lakes, groundwater, subterranean aquifers, the polar icecaps, and saturated soil (tundra or wetlands). The cycle is the process of transferring water from one state or reservoir to another through gravity or the application of solar energy, over periods that range from hours to thousands of years.[10] The whole system works only because more water evaporates from the oceans than returns to it directly in rain or snow. The balance falls as rain or snow on land, and it is this balance that makes our lives possible, for when the rain falls, it falls as fresh water. There is not only quantitative but also qualitative renewal: the process scrubs the water of its impurities and delivers potable, usable water to the biota, which includes us.

For obvious reasons, most evaporation is from the ocean. Willie Sutton robbed banks "because that's where the money is," and the oceans yield up most evaporation because that's where most of the water is. But evaporation also takes place from lakes and rivers, soil, and even snow and ice, in which case it is called, for reasons now obscure, sublimation. Plants also exhale water; the evaporation of water through minute pores, or stomata, in the leaves of plants is called transpiration. Most hydrologists simply lump transpiration, sublimation, and evaporation together and call it evapotranspiration.[11]

The best estimate among the many educated guesses — the estimate of Igor Shiklomanov and his State Hydrological Institute in St. Petersburg — is that there are some 1.4 billion cubic kilometers of water on earth, in liquid and frozen form, in the oceans, lakes, streams, glaciers, and groundwater. And even Shiklomanov, a formidable figure in the water world, and the man the United Nations selected to do its world inventory of water resources, suggests that this is a crude guess: no one really knows how much water is stored in underground ice in permafrost regions, for instance, or in bogs and marshes. His own estimates on these matters, he admitted to a UNESCO conference, were based on numbers arrived at "computationally, under fairly crude assumptions." The amount of ice on earth is, after all, constantly changing. About 18,000 years ago, ice covered one-third of the earth's surface. Now, only about 12 percent remains icebound, and that proportion is shrinking. If global warming is real, it will shrink more and faster.

Some 1.4 billion cubic kilometers: how much water is that? The number is so huge it's hard to grasp. How to put this into any kind of human perspective? A common metaphor, useless but often trotted out by water statisticians, is that "spread evenly, it would cover the earth to a depth of 2.7 kilometers." All of which means what? How many cubic meters is the average iceberg? The average lake? The average river?

Hardly any of this 1.4 billion cubic kilometers is useful for human consumption. More than 97 percent is ocean water, too salty to drink or to use for irrigation. Freshwater stocks are only 2.5 percent of the total, and spreading that amount evenly over the globe would make a skin a mere 70 meters deep.

But even this 2.5 percent of the water supply isn't all usable. A trivial amount is in the air at any one time in the form of rain, fog, or clouds, about 0.001 percent of the total. An even more trivial amount is in the "biosphere" — in us and other living things, including plants — about 0.00004 percent. But really significant amounts, slightly more than two-thirds of the total, or 24 million cubic kilometers, are locked into polar icecaps and permanent snow cover. And a large percentage of the remaining 16 million cubic kilometers lies too far underground to exploit, imprisoned in the pores of sedimentary rock.

Freshwater lakes and rivers, which are where humans get their usable water, contain only about 90,000 cubic kilometers, or 0.26 percent of the world's total supply of fresh water, which is itself only 3 percent of global water supplies. Put another way, if all the earth's water were stored in a 5-liter container, available fresh water would not quite fill a teaspoon. Or, to use the metaphor of spreading the stuff evenly over the globe, available fresh water would make a layer only 1.82 meters deep. A really tall person, say a basketball player who couldn't swim, could stand upright in that and not drown.

Fresh water is renewable, at least in the sense that the hydrologic cycle evaporates water from the oceans and returns a good deal of it to the land. This water eventually makes its way back to the oceans through rivers, streams, lakes, and underground aquifers. An enormous amount of water evaporates from the land and oceans annually, using up about half the solar radiation reaching earth. Shiklomanov, as usual, has found his computer equal to the task and has calculated the amount at about 505,000 cubic kilometers. Other estimates put it higher, at nearly 575,000 cubic kilometers. Looked at another way, the top 1.4 meters of the sea evaporates every year.

The amount of time that water stays put in any one place is called its "residence time." Residence times vary tremendously, from ten days for the atmosphere to somewhere around 37,000 years for the sea. Lakes, rivers, ice, and groundwater have residence times lying somewhere in the middle and are highly variable. Most rivers renew themselves completely quite rapidly, in about sixteen days. Groundwater and the largest lakes and glaciers can take hundreds or even thousands of years. If the Upper Island Cove bergy bit really was 12,000 years old, that would put it near the middle of the range.

Almost all the half-million cubic kilometers of evaporated water, around 458,000 cubic kilometers, or 90 percent by Shiklomanov's measure, falls back into the sea as rain or snow. But

since evaporation exceeds precipitation on the seas, the net difference, somewhere around 45,000 or 50,000 cubic kilometers, falls
on otherwise dry land. Another 60,000 cubic kilometers is rain
and snow of purely local, or non-ocean, origins. About one-
third of all the precipitation falling on land — somewhere around
34,000 cubic kilometers — goes back to the oceans in rivers and
groundwater runoff.

"Runoff" is the renewable aspect of water resources, the dynamic part of long-term water reserves, an index, if you like, of
viable water supply. And so at last we get to the human community's usable freshwater supply: about 34,000 cubic kilometers a
year.

Humans are already using more than half of it — 35 percent for
irrigation, industry, and households, and another 19 percent to
meet instream needs. It sounds as if there is capacity to spare, but
the remaining half is the hardest — and most expensive — to ac-
quire. All the easiest aquifers have been tapped, the easiest rivers
dammed. The population is still increasing at an alarming rate.
The ecological costs of using all the water in any system have become only too apparent. Water demand tripled between 1950 and
1990, and was expected to double again in thirty-five years. Where
is this water to come from?[12]

Still, that available water, that 34,000 cubic kilometers, would be
enough to supply every human on the planet with about 8,000 cubic meters a year — ample by any measure — if it were distributed
evenly. But it isn't distributed evenly. There are places that don't
need it that have too much, and places that desperately need it that
haven't nearly enough.

Twenty percent of global runoff comes from the Amazon Basin
alone, while some parts of South America are the driest on earth:
the rain gauge of Arica, in Chile, routinely records zero annual
precipitation, and in the first half of the twentieth century did so
for forty years straight. In the Sahara, the greatest desert on earth,
there are virtually never clouds. On the wet end of the scale,

Mount Waialeale in Hawaii has recorded more than 11.5 meters of rainfall — almost 40 feet — in a single year.

The twenty-eight largest freshwater lakes in the world, over-whelmingly concentrated in northern regions where glaciation has scored deep holes in the earth's crust, account for 85 percent of the volume of all lakes worldwide. Lake Baikal in Russia alone ac-counts for one-quarter of all the world's lake-held fresh water (23,000 cubic kilometers). Africa's Lake Tanganyika is second in volume (19,000 cubic kilometers), and Lake Superior, on the U.S.-Canadian border, is third at 12,000 cubic kilometers. The North American Great Lakes, the world's largest lake system, account for 27 percent of global lake volumes.

Of the twenty-five largest rivers of the world, three are in Africa (the Congo, Niger, and Nile, with a combined runoff of 1,982 cu-bic kilometers); four are in South America (the Amazon, Paraná, Orinoco, and Magdalena, with a combined runoff of 8,829 cubic kilometers); eleven are in Asia (the Ganges, Yangtze, Yenisei, Lena, Mekong, Irrawaddy, Ob, Chu Chiang, Amur, Indus, and Salween, with a flow of 5,722 cubic kilometers); five are in North America (the Mississippi, St. Lawrence, Mackenzie, Columbia, and Yukon, with a combined flow of 1,843 cubic kilometers); and only two are in Europe (the Danube and the Volga, with a combined flow of 468 cubic kilometers).

In national terms, Brazil has the most water, containing one-fifth of all global resources. The various countries of the former Soviet Union are collectively second, at 10.6 percent of global fresh water. China (5.7 percent) and Canada (5.6 percent) are third and fourth.

China and Canada have virtually identical resources, but China's population is thirty times greater. Water cannot be counted in isolation to human need and numbers.

Nor can water be assessed in isolation from its other, non-human purposes. Even if we could, we wouldn't shift Brazil's 20 percent to, say, the Sahara. Doing so would put an end to the great-est reservoir of biomass on earth, and the planet's greatest rain forest, which we have come to understand is its respiratory system. To do so would be like placing a giant vise around the earth's lungs.

The human use of the runoff that reaches the sea has little effect on the oceans or their saline balance, except for the imponderables of pollution, and except in small enclosed seas such as the Black, the Baltic, and the Mediterranean. But the diversion and use of streams that would otherwise enter hydrologically closed systems can fundamentally change their water-salt balance and lead to ecological catastrophes, as we shall see.

There is another problem, one of timing and erratic supply. In monsoon regions, virtually all the runoff comes in a few short months, often causing severe flooding. In the southwestern United States there is virtually no rain from May to September, when it is needed most, an imbalance that has conjured into being one of the grandest follies in all human history — the irrigation of what are, essentially, deserts. Water policy, and the avoidance of conflicts over water, require that these imbalances be managed. Often, different areas within a single ecosystem or river basin will have wildly variant needs and demands. Some are badly water stressed, while others have a surplus. It is in these transnational and transregional systems that the potential for water wars continues to exist.

I grew up in an arid zone, the austere but beautiful landscape north of the Great Karoo in South Africa, on the approaches to what my ancestors called The Thirstland. The rivers there seldom flowed. For most of the year they were reduced to muddy puddles, and sometimes these, too, vanished. My grandfather was a farmer in this uncertain land, and water — the idea of water, the longing for water — consumed his life. With sweat, fanatic labor, and the help of teams of oxen, he and his workers built a number of mud dams, but sometimes years would go by with no rain whatsoever, and the dams would shrink and his livestock perish. When it did rain, it rained in cloudbursts, and he would patrol his dams gloomily, watching for faults that would tear the retaining walls apart.

Toward the end of his life, my grandfather brought in a drilling rig and sank a borehole 300 meters into the rock and shale of the substrata. At around 250 meters the drillers found an aquifer, and although I was just a child I still remember the cold, clear drops of water from the center of the world mingling in the dusty earth of the farm with the old man's tears. To my grandfather it was a miracle, a purely good thing, life extracted from rock.

In this modern, more highly populated, and less optimistic age, what he was doing is less well regarded. It is called "water mining," which is the unsustainable withdrawal of water from an aquifer that is no longer being "recharged" or refilled. My grandfather operated in a sparsely populated part of the world, but if there had been enough people doing what he did, the aquifer would have begun to shrink, and eventually it would have run dry and all farming would have had to cease. This, too, may seem like a small thing in a faraway place, but aquifer depletion is happening in many more difficult and important places in the world, in many parts of the United States (particularly in the Southwest and the High Plains), in Mexico, in the Middle East, in India, in many parts of the former Soviet Union, in northeastern China, and in southern Europe. Groundwater depletion is one of the great unseen but looming crises facing our planet, with all its implications for diminishing food supplies, for human misery, famine, conflict, and war.

Late in 1998 Ed Ayres, the editor of *WorldWatch*, the monthly journal of the WorldWatch Institute, asked a critical question. He had been noticing — as had anyone who cared about water — the reports about falling water tables coming in from all over the globe, most disturbingly in major food production areas. In northeast China water tables had been dropping by more than 1 meter a year. In the Punjab the drop had been more than that. The water table throughout the Mexico City region, notoriously, had dropped 20 meters in fifty years, causing large parts of the city to subside. In California's Central Valley there had been a similar though smaller subsidence. A quarter of irrigated cropland in the United States was experiencing drops — from 20 centimeters to more than 1 meter a year. Some irrigation wells had ceased to draw water at 300 meters and were now down to almost 1,000 meters. Everyone, Ayres noticed, was tracking the falling water levels, but no one seemed to be asking the critical question: If the top of the aquifers is dropping, where's the bottom? The answer, he found very often, was: We don't know. "Mostly, it seems, we find out where the bottom is when a large number of farmers' fields go dry," he said. "If the people taking out the water aren't anticipating when they're going to hit bottom, they're in for a staggering shock."

Ayres was professionally concerned with the environment, and he had a larger point to make:

> In our general distraction with the complications of managing even the here and now, we forget to focus on determining at what point that shock will come, and what we will do then. We know for example that you can't go on killing off the species in an ecosystem indefinitely; at some point [it] will fail like a building in which a key beam gives way and the whole system collapses. Yet we keep pushing the limits . . . we keep forgetting about that inevitable point of shock where the water supply is abruptly gone — where the lake goes anoxic, the forest turns to desert, or the river turns to dust.[13]

The process of taking more water out of an aquifer than naturally returns to it is called "groundwater overdrafting." It is a modern phenomenon, depending on reliable pumps and cheap energy. Before mechanical pumps, societies might have built dams and irrigation canals and aqueducts, but they were dependent on gravity for directing the water to its intended destination. Cheap oil changed all that. Among other things, it made deep wells possible, and these deep wells exploit underground aquifers that have, in some cases, slowly accumulated for thousands of years. The limits are clear, both environmental and economic: either the water runs out, or else the aquifer is so deep that even cheap energy at some point becomes too expensive to pump the water to the surface. At that point the pumps are turned off and the users are forced to move or to find alternative sources of water. Sometimes this is not possible, and the farming just stops.

There are other problems with falling water tables. They may cause rivers to dry up. And in many critical cases the withdrawal of too much water in coastal aquifers causes seawater to intrude, ruining the aquifer permanently for human use.

Like surface water (lakes and rivers), subsurface water, usually called groundwater, is constantly on the move. Over the years or decades or centuries it eventually finds its way into the channels of rivers and streams and then back to the sea. It, too, is replenished through rain. Somewhere around 10 to 20 percent of rain finds its way into underground water systems; in return, groundwater contributes, on a global basis, about 30 percent of total river runoff. It has a stabilizing effect on rivers, minimizing the differences between wet and dry seasons. The movement is much slower than runoff on the surface, but it does move.

The process is well understood, though its measurement is still haphazard. Rainfall — at least that part of it that doesn't directly join free-flowing surface water or become absorbed by plants — seeps into the soil zone, the upper meter or so of earth. At first there is a "zone of aeration," in which the small interstices between soil particles are filled with a mixture of water and air; if the precipitation continues, complete saturation of the soil zone occurs and the water continues to descend until, at some point, it merges into a zone of dense rock.[14]

Some rocks are too dense to allow penetration by water; these impermeable rocks are called aquicludes. Others are more porous and store considerable amounts of water. These rocks are called aquifers. The geology can get quite complicated. Aquifers can sometimes lie underneath layers of aquicludes; these are called confined aquifers. In other cases, called unconfined aquifers, there is nothing around the aquifer but unsaturated and permeable material such as gravel, shale, or sand. The boundary between the unsaturated stuff and the water-bearing rock is called the water table. In water-rich places such as Canada, northern Europe, and parts of the United States, the water table might be relatively close to the surface. Depending on the subsurface geology, the aquifer underlying the water table might be several hundred meters deep, in others only a few meters. In certain arid zones, where the aquifers are either not being replenished or being replenished very slowly, the water table can be hundreds of meters below the surface. Where no permeable rocks exist, or where they exist but climate changes have allowed no recharge, there is no water table at all. In other cases, water rises without pumping; these "springs" happen in confined aquifers where the hydraulic potentials are greater than the energy required to bring the water to the surface, and where there are small natural fissures. A well drilled into one of these pressurized aquifers will cause the water to gush to the surface, and it will continue to gush until the pressure has been equalized. Some of these "artesian wells" have been lifting water to the surface for decades. Eventually, though, the pressure will equalize, the potentials will decline, and the well owner will have to resort to pumping.[15]

Groundwater is tracked by remote sensing and tracer techniques, but its movement is exceedingly difficult to follow. It is known that groundwater migrates slowly, sometimes only a few millimeters a day, though occasionally a few meters. Near the water table the average cycling time of water may be a year or less, while in deep aquifers it may be as long as thousands of years. It's easier to measure water tables. Through test wells and controlled pumping, it's not difficult to measure the recharge rate and flow behavior around a particular site. The difficulty comes in sensing movement in the aquifer as a whole. Water can be stored in the pores of rocks, but it can also be stored in cracks and fractures, and sometimes these fissures can cause water to travel quickly and over great distances. Sometimes these "preferred pathways" become subterranean streams or even rivers, which can help swell surface rivers dramatically during heavy rains. That is why aquifers are almost impossible to clean up after they have become polluted, and why pollutants, leaching through the soils from waste dumps, landfills, and agriculture, can occasionally be found at alarmingly long distances from the polluting source.

No one knows how many unexplored and unexploited aquifers exist, but the amount of water stored in them is thought to be considerable. Water exploration companies have not yet made a major impact on the stock markets of developed countries, but they wouldn't be a bad bet for adventurous investors.

The slow replenishment of groundwater is called "groundwater recharge." It happens, obviously enough, mostly during the rainy season, or during the winter in temperate climates.

When the water table is deep underground, the water of an aquifer may be exceedingly old, sometimes left over from a previous climatic epoch. The best-known example is the Nubian sandstone aquifer, which underlies what is now the Sahara Desert, but which was even in historical times (maybe eight millennia ago) verdant grasslands and occasional swamps; fish and hippopotamus fossils have been found in many parts of the Sahara. Radioisotope dating techniques have shown that this water is many thousands of years old. This ancient aquifer is being exploited, as we shall see, by Libya's Muammar al-Qaddafi in the planet's most grandiose episode of water mining.

Colonel Qaddafi's, however, is far from the only instance of water mining. The Ogallala Aquifer, a 100-meter-thick band of sandy material that underlies a number of the Great Plains states in the United States from South Dakota to Texas, has been mined for years and is rapidly being depleted. Water equivalent to about one-third of the volume of Lake Huron has already been extracted from this largest of fossil aquifers, at a rate far greater than that of recharge. Perhaps 60 percent of it, some 6 billion tons, is already gone, and substitutes are not readily available.

Many other arid and semiarid zones are similarly mining their water. Fossil aquifers are not affected by short surface droughts, and they are found in areas where there is no other reasonable supply of water. This is why Ed Ayres was worried. At some point these mines, like all mines, will exhaust the resource. They are almost all being used for water to produce food. And the world's population is still growing. When the water is exhausted, what then?

I have asked myself many times what my grandfather would have said if he had known about Colonel Qaddafi, about the desert cities tapping into the Ogallala, about the whole notion of unsustainability. He was a large man, weighing nearly 140 kilos, or 300 pounds, when he died, and I remember him best sitting on the stoop of his little farmhouse, drinking large mugs of strong black coffee, staring down the row of bluegum eucalyptus trees into the dusty, arid distance. I loved him fiercely, as only a grandchild can, and was never happier than when I was trotting at his side, visiting his small and usually empty dams, his sheep-shearing sheds, the pens where the milk cows were kept. Afterward I would sit there on the stoop with him and he would let me sip from his mug, and together we would watch the distant dust clouds march across the horizon as a neighbor's cart passed down the sandy rural roads. A few hundred meters from the farmhouse was a round corrugated iron storage tank, and above it a windmill clanked, chugging water up from the reservoirs far below. My grandfather was a man who understood the soil and knew how natural systems worked, and he probably had an intuitive understanding of how that water had been secreted away in the interstices of the subterranean rocks, so many centuries before. He would have considered it a gift from God, I think, to be used wisely and not squandered, but in any case used as you use the rain that is also given by God, to produce things as farmers must. He never knew of its frightening finiteness. And thus are ecosystems ever changed, more by the cumulative deeds of good men doing what they believe in than by the rapacious actions of Green demonology.

My grandfather had enough troubles in his life. I'm glad he never had to wrestle with this one.

3

Water in History

Some things never change: how humans have always discovered, diverted, accumulated, regulated, hoarded, and misused water

THE ANCIENTS may have seen God everywhere, but they also saw opportunity. Superstition was often accompanied by a politics both cynical and wise, and by engineering skills of a very high order. Water has been used to nurture cultures, but also as a weapon of war, and as a convenient cover for crimes best hidden. Of course, nothing in that prevented ancient cultures from withering away, victims of hubris, or drought, or, most likely, both.

In modern times, along with the often conflicting notions of the rights of the individual and the sovereign nation-state came the need for water law. This is a saga still unfolding.

I find this passage in my notes, cobbled together from various texts over the years: "Cultivation terraces, some of them more than 3,000 years old, can be found in many parts of Asia Minor, Europe, Africa, Asia, and Andean America. The earliest urban societies all depended on systems of dikes, canals, and aqueducts, many of them designed and built to exploit and control seasonal flooding. Some of these were significant in size: a dam built below what is today Cairo somewhere in the third millennium B.C., at Sadd-el-Kafara, was estimated to contain a hundred thousand tons of earth and rock. There are qanats [horizontal wells] in the Iranian desert three thousand years old, some of them still functioning."

I once ran into an archaeologist in Tanzania who wrote prose like that — workmanlike, sturdy, but essentially dull. He knew it, too, and despaired of it, because what he really wanted was to communicate his passion for the ancient lives he found in the runes and ruins of his field trips, and all he could produce was prose that would make graduate students' eyes droop. I found him in his tent, perched on the stony floor of the Great Rift Valley, near the Ngorongoro Crater, and close to the Olduvai Gorge, where the famous Leakey family began unraveling the story of the human species. This is one of the most eerily evocative places on earth. Nearby, impressed in the mud many millions of years ago, are the Laetoli footprints, fossilized imprints of two adults and a child, our earliest ancestors, left there for the contemplation of their distant descendants. The archaeologist, whose name was Gerrit Witman, had been in the area for more than seven years, funded by his university in Germany and by the grants he could scrounge from American foundations. When I came across him, his head was in his hands and he was staring gloomily at a pad of paper in front of him. The page was blank. I started to laugh. A blank page is something all writers know only too well.

He said nothing for a while, then he suddenly sprang to his feet and dashed outside. "Come with me," he said, and strode off into the bush.

I looked at the man who had brought me here, a Swahili speaker from the coast near Dar-es-Salaam, but he just shrugged and raised his eyebrows. "Might as well," he said, and set off in pursuit.

We caught up with Witman at the foot of a stony hillside. He was perched on a boulder, teetering a little.

"What is it?" I asked. "What do you see?"

He pointed to the left, where the hill flattened into the plain. "Over there," he said, "about fifty miles, that's Lake Manyara. Up from there is the Mosquito River. It's new, I think. It didn't exist centuries ago. There were no rivers here, but there was water up on the plateau, there." He shifted a few degrees on his boulder, chopped his arm upward, pointing at a crest we couldn't see. So what were people doing farming here?"

"They were?"

He scrambled up the hillside a hundred meters or so, calling for us to follow. Near the top, he stopped. "Look. See . . ." He pointed to his feet. He was standing in a shallow gully, which ran arrow-straight across the face of the hillside, with a gentle downslope. "These are irrigation trenches. At the bottom of the hillside there were small fields with stone borders where sorghum was grown. The neatness! The precision! Some of those fields are immense, but some of them are less than a meter square! Think of the kinship patterns and the needs that drove them to carve up the land in ever-smaller parcels. This is an irrigation system. There are better preserved ones further up, but these are far, far older; we haven't yet worked out how old."

"So what's the problem?" I asked.

"The problem," he said, "is that there were cultivators here, possibly a millennium ago or more, in this arid place, which was arid even then, a culture with villages, laws, government, chiefs . . . What I mean is, it's so exciting, imagining these people, these vanished and long-gone people, and putting yourself into their heads and thoughts, imagining their lives . . . And then you write it down, and it looks so dull . . . 'So there are stone channels in the hills?' people say; 'big deal, these are no pyramids.'"

"So you're writing for students and scholars, not for a screenplay," I said, a little unsympathetic. Indeed, I admired him for his imagination, but not, when I saw it later, for his prose.

I went back to look at those stone channels built by that long-ago people. Not pyramids, no, but the engineering skills were far from insignificant. I tried to imagine those men — and women, this being Africa, where the women did most of the work — toiling in the long ago, rolling these stones up the hill, trenching, terracing. Who had been the overseer, the person with the vision, the surveyor in chief? What kind of person could imagine taking water to a place where there was none? Farther up the Rift, another archaeologist, John Sutton, had found many more of these channels, engineered responses to water needs emerging spontaneously where they were needed. Sutton was working in similar countryside. There, too, the surroundings were too dry for agriculture, but a few rivers rose on the lusher Crater Highlands, and an unknown number of centuries ago these rivers were also harnessed and channeled into even longer and better-preserved stone-built canals, cunningly embanked and leveled and meticulously maintained, some of them more than 5 kilometers long, ending at tiny stone-enclosed fields covering more than 2,000 hectares.[1] Like Witman's, this unknown culture, here and at other sites by lakes Eyasi, Manyara, and Natron, vanished, abandoned for reasons unknown — perhaps because the nomadic Maasai were moving into the area, perhaps because increasing population had overstrained resources, or perhaps because deforestation for firewood had critically wounded the streams, causing them to dry up.

I thought back to my notions of who had built these channels. I admitted that I was prone to my culture's failure of imagination. Somehow, we have lost a real sense of history. We have come to hold the quaint notion that because people lived long ago, they must perforce have been more "primitive," more driven by superstition than we are, and we are constantly surprised at evidence of planning, forethought, ingenuity, and skill. I thought then that it ill behooves a culture in which large numbers of citizens believe in abduction by aliens (and in which a woman was recently charged with attempted murder in a U.S. court for hiring a fortuneteller to put a spell on an enemy) to pour scorn on the arcane beliefs of ancient cultures, and even less so when it is clear that folklore cohabited comfortably with engineering and management skills of a very high order. More than 2,400 years ago Darius the Great of Persia, a ruler noted for his monumental building projects and great administrative skills, encouraged the construction of new water sources by exempting from taxation "for five generations" the revenue derived from the building of new qanats, an early instance of government pump priming of the economy by subsidizing entrepreneurs. *Plus ça change* and all that. A primary solution for modern water problems is to treat water as an economic good; to Darius, this was already plain good sense.

Still, it is interesting what folklore has grown up around water. Water is so necessary, so central to all life, that it isn't surprising it took on a significance beyond the intrinsic. In a world where gods inhabited everything, gods most certainly inhabited water. And it is always interesting how politics and religion intersected where it came to water.

> *Spring up, O well,*
> *Sing ye to it*
> *Thou well dug by princes*
> *Sunk by the nobles of the people*
> *With the sceptre, with their staves*
> *Out of the desert, a gift. . . .*

This biblical verse refers to a custom in which a new well is lightly covered over and then miraculously uncovered in the presence of the tribe's leaders, thereby assigning it as the property of the whole clan. In modern times we'd be more cynical. We'd call this ceremony a photo op and impute baseness to the politicians who orchestrated it. The song served as a kind of ribbon-cutting ceremony. In either case, the community ended up with a benefit it hadn't had before.

The peoples of Mesopotamia and the Middle East, water stressed even in biblical times, have water tales running all through the Word. In the Book of the Prophet, believers are enjoined to share water with whoever needs it, a basic part of the human obligation toward strangers. Did not the world, after all, result when God divided the waters of the deep from the waters of the air, and so created the earth? In the Koran the Lord says, "We clave the heavens and earth asunder, and by means of water we gave life to everything." God is the fountain of living water. The bean counters of exegesis say that there are more than two hundred references to water or wells or oases in the Bible, and I see no reason to doubt them. Doesn't Moses mean "drawn from the water" in Hebrew?

The Gnostics regarded water as the primary element, and their influence was felt among the Jews. Judah ben Pazi, passing on the wisdom of Rabbi Ismael, had the wise man saying: "In the beginning the world consisted of water within water; the water was then changed into ice and again transformed by God into earth. The earth itself however rests upon the waters, and the waters on the clouds."

The Well of Miriam, the source of life for the Jewish wanderers in the Sinai, is said to be located in the deeps of the Sea of Galilee (which might, in truth, once have stretched through Eilat to the Dead Sea). The mystic of the Kabbalah, Isaac Luria, felt that the water of the Well of Miriam was one of the centers of cosmic energy; anyone who immersed himself in these waters would understand the mysteries, and his memory would never fail.

Muslims believe that the Dome of the Rock in Jerusalem is closer to God than any other place on earth, and that beneath it originate all the sweet waters of the planet. The ancient cultures had the same notions. The Egyptian god Hapi, a male, was often shown with two full breasts, from one of which flowed the northern Nile, the other the southern. Nun (or Nu) is Chaos, the primordial ocean, the germ of all things and all beings. He was the "father of the gods" and is sometimes found represented as a personage plunged up to his waist in water, holding up his arms to support the gods who have issued from him.[2] The Sumerian god Aspu was the governor of the Sweet Waters; Ea of Mesopotamia, the god of water, set up his palace in the Fertile Crescent where the Euphrates and the Tigris converge; Gilgamesh rooted the Tree of Life in the same place. The Babylonian god also called Nun personified the idea that water was the source of all life, that historically the earth came forth from the water, and that water was the quickening element of all creation. "I shall shut up the heavens, and no rain will fall" — always the curse of a vindictive and cruel god.

Around Accra in western Africa, the pond spirits were locally powerful. They must be particularly irritable these days, since the modern Ghanaian capital doesn't treat its sewage but dumps it into dry wells to seep away; in the rainy season the wells overflow and pollute the city and the groundwater.

Among the Dogon of Mali, whose farmers are famously creative with the meager water they receive, water was a practical matter but also integral to the creation myth, in which water is a divine green seed that impregnates the earth so that it brings forth twin green beings, half man, half serpent. Humans originated when the supreme celestial god, Amma, created in his likeness Nummo, the god of "wet" water. Water, the life principle, is closely involved in the rites accompanying childbirth. When the afterbirth is expelled, proving that the child has indeed been born, one of the midwives takes a mouthful of water and sprays it gently over the baby. The cool touch of the water makes it cry out — by which the child has "officially" received the gift of speech.[3]

Lest we be tempted to believe that these were just quaint primitive customs — well, they were, but they were just as common among the quaint primitive Europeans as anywhere else. Water worship and holy wells survived in England long after the Druids were a distant folk memory. Pope Gregory I, deliberating on the English in a letter to Abbot Mellitus in the year 601, recommended countering the water superstitions of the locals not by destroying their places of worship but by subsuming them by sprinkling "holy water upon said temples . . . that they might be converted from the worship of demons to the worship of the true God." But the old ways often persisted. An old book of folklore records the supplication of a Scottish peasant: "O lord, Thou knowest that well would it be for me this day an I had stoopit my knees and my heart before Thee in spirit and in truth as I have often stoopit them after this well. But we maun keep the customs of our fathers."[4] In 1893 R. C. Hope published *The Legendary Lore of the*

Holy Wells of England, in which he catalogued the names of 129 saints to whom anciently venerated wells had been transferred, the Holy Virgin being the most common, closely followed by the ever popular Saint Helen, the tabloid heroine of her time (she was the English-born mother of Constantine, the first Christian emperor).

The most famous holy well in England is at Holywell in Wales. Winefride, the beautiful daughter of the chieftain Thewith, resisted Prince Caradoc's advances after she had dedicated her life to virginity and God. He caught up with her at the top of a hill and cut off her head, but after it rolled to the foot of the hill, a copious stream burst forth, forming a well that was lined with fragrant moss, the stones at its foot "tinctured with the blood of the youthful martyr." Winefride was subsequently miraculously reunited with her head and became a nun, dying in 660.

The indefatigable Brothers Grimm recorded many instances of well worship along the Rhine Basin. "Sacred wells and fountains," they wrote in *Teutonic Mythology,* "were rechristened after saints, to whom their sanctity was transferred." Ancient customs and rituals survived, among them oaths and ordeals, consecrations and processions, but clothed in Christian forms. In some customs there was little to change. "The heathen practice of sprinkling a new born babe with water closely resembled Christian baptism."

As in other endeavors, religion blends with pseudo-science, itself developed as an explanation for the unexplainable. In arid regions in ancient times, the mere finding of water was high art. Water divination was a valued practice, and water diviners ("dowsers") have existed in all cultures at all times. They are still active in the West, and rural people in search of a well commonly call on such a one, trolling across a field holding a forked piece of wood, or even a coat hanger bent into the right shape. Like many urbanites I've watched them work, not knowing enough even to be skeptical; maybe there is some principle at work here, ill understood but efficacious. Knowing where to drill for water is a serious business in South Africa's Karoo region, for example, because water is scarce there and the drills must go very deep. I once watched an old Boer patrolling a hectare or two, waiting for the twinge that showed him where the aquifer was to be found. He was a grand old fellow, bearded like Abraham the Prophet, but he always wore baggy khaki shorts and knee socks, which made him look like a wrinkled old Boy Scout. He used a yew fork for his work, imported from England; he claimed that yew knew water better than any other wood, and no one dared to argue. It took him most of the day, criss-crossing the homestead a hundred times before he was satisfied, but when the drill went in, it found water. No one was ever surprised — everyone knew that water dowsing worked. It was a natural law, like the monsoons and man's venality.

Superstition — and engineering. Didn't Solomon say, "I made pools of water to irrigate a forest springing up with trees. . . . Thus I gained more wealth than anyone else before me in Jerusalem"? It was in Babylon, of course, that Nebuchadrezzar II built his Hanging Gardens to please his Median wife, Amytis, who was pining for her verdant homeland. What she got was on everyone's Seven Wonders of the World list. They were described by the ancients with suitable awe: a massive series of terraced roof gardens, each lined with layered reeds, bitumen, and lead to prevent seepage to the levels below. The gardens were irrigated with pumps from the Euphrates River and from a deep well, the water hauled to the top by a chain pump.

The digging of shallow wells was the earliest organized change from the communal waterhole. Wells went deeper as the need (that is, the population) grew and as the tools were developed. Well drilling went astonishingly deep astonishingly early — as early as Neolithic times. The Egyptians had perfected core drilling in stone quarries by 3000 B.C., although the wells of the time, of large diameter, lined with stone and constructed with human and donkey labor, rarely went much below 50 meters. In China, however, a churn drill was developed two millennia ago, and some of the wells these devices produced went down more than 1 kilometer into the soil and rock, though they were merely made of wood and powered by humans, and sometimes took years, even decades, of unremitting labor to finish.

Europeans, cut off from the civilizations to the east and their engineering skills, came relatively late to well drilling, perhaps because water was relatively easy to find in their temperate climate. In the early twelfth century, "flowing wells" were discovered in Flanders and England. In the same century percussion drills were invented; monks made them, as they had introduced so many other innovations, such as flow-through toilets and in-house sanitation. Carthusian monks from Lillers in France drilled an early well in 1126, a thing of much marvel. A few years later other religious near Gonnehem in the province of Artois drilled four wells 100 meters into the fractured chalk below; they gushed so strongly that they drove a water mill nearly 4 meters above ground, and flowing wells have been called "artesian" after these Artois wells ever since.[5]

By 1340, Bruges in Flanders had a municipal water distribution system, a central cistern from which water was pumped by buckets on a chain underground to public faucets. Two centuries later, London set up an awkward contraption under London Bridge, a series of five waterwheel-driven pumps, which supplied the whole city with water.

The earliest irrigation schemes were called qanats, which were both well and aqueduct. Qanats made the great urban civilizations of Mesopotamia possible, and they are still widely used, from Afghanistan through Iraq and Iran all the way west to Egypt. A qanat is essentially a horizontal well. Aquifers, groundwater close to the surface, are found on uplands, often the foothills of mountains. A shaft is bored horizontally into this alluvial fan, which is usually fed by mountain streams. The tunnel is constructed to slope gradually downward, so the water flows entirely by gravity to its destination, which in some of the more remarkable old qanats can be 50 kilometers away.

If you fly over the desert in southern Iran, you can easily track the course of the longest qanats. Every few hundred meters, for kilometers across the stony plains, you can plainly see a mound of earth surrounding a hole bored into the soil, each hole a vertical shaft down to the qanat tunnel itself. These rows of manholes, for that's what they are, have multiple functions. Their primary purpose is to serve as a means of extracting the rubble and soil from the excavation of the tunnel itself. They are also access points for repairs, breathers to aid circulation and water purity, and sources of irrigation water. At the lip of many a manhole there are still rickety contraptions of poles and rope, homemade winches. The qanat itself can be many meters below the surface, the water protected from evaporating in the desert sun. Many of these qanats have been operating for thousands of years; until the building of catchment reservoirs in the 1930s, Teheran got all its water from a dozen qanats that put out something like 800 liters a second. Because most Iranian towns of any consequence are supplied by qanat, they have been built close to mountains and have a canal running down a main street.

The ancients were also shrewd enough to know that it was not enough for water just to be water. It had to be clean water. Texts dating from as early as 2000 B.C. outline techniques for cleaning water through sand filtration and boiling. A guide to the supposed tomb of the apostle Saint John in Compostela, Spain, made sure that pilgrims of the eleventh and twelfth centuries knew "which waters were fair, and which foul, which were fit to drink, and which not fit for asses."

The Romans were the greatest builders of water distribution systems in the ancient world, but they learned their techniques from the qanats of the Fertile Crescent and from the Assyrians, who three millennia ago used more than 2 million huge blocks of limestone to construct an arched aqueduct to bring fresh water to the city of Nineveh. The Assyrians were inventive engineers; the water was carried across a river valley in a limestone channel 10 meters above grade and 1 kilometer long. Even more brilliantly, they managed to construct an inverted siphon to increase the flow into the aqueduct, a feat not duplicated until the nineteenth century. Less positively, the Assyrian king Sennacherib burned Babylon to the ground in 695 B.C. and diverted a major irrigation canal so that its waters would cover the ruins, hiding the crime. The Assyrians themselves were said to have learned their techniques from their predecessors, the Sumerians.

Most of the Roman aqueducts were built in a spurt of construction between A.D. 312 and 455, and remnants can still be seen in Italy, Greece, France, Spain, Syria, Morocco, and the old Persian Empire; some of them are still in use. The Roman system, described in detail by Sextus Julius Frontinus, the supervisor of the empire's waterworks, used eleven major aqueducts to bring water more than 40 kilometers to the city in sinuous, curving channels that were themselves almost 100 kilometers long, mostly in underground tunnels made of stone, terra cotta, and a variety of other materials, including brass, wood, and, notoriously, lead — an early instance of inadvertent industrial pollution. The system was gravity fed and flow-through, like the latrines of Rivault Abbey in Yorkshire and the trendy toilets of California; the surplus water was used to power the city's fountains and to flush its sewers into the Tiber.

No one improved on this Roman delivery network until the industrial era, when the Americans invented pipes that could sustain pumping under pressure. These pipes were made of asphalt-treated wood, banded with iron. For those interested in industrial archaeology, pipes like these still take water to the oceanside villages of Witless Bay in Newfoundland.

All these developments pale, of course, before what the Californians have done, which is to spend billions of dollars diverting entire rivers, sometimes pumping them over a water basin and a continental divide, to bring a cultivated civilization to a place where it should not be. Their network is still not finished, and perhaps never will be (for it grows ever more grandiose), but it already brings in water some 900 kilometers, or 550 miles, with an annual yield of over 4 million acre-feet (there are 1,234 cubic meters, or

1,234 tons, per acre-foot). The Californians have made the desert bloom — but at what cost in money, resources, and impact on the environment they are only just beginning to find out.

Irrigation underpinned most of the ancient cultures whose imprint has been left to us. As we shall see, irrigation carries its own dangers, and many irrigating cultures perished through their very success, expanding until they exhausted the resources on which their success was based. One of the interesting things about ancient water-dependent civilizations — cultures dependent on irrigated agriculture — is why they collapsed. Why did the cultivators of Tanzania who so tormented archaeologist Witman vanish? Why did the Sumerian culture disappear, to be replaced by the Assyrian and then by the Babylonian cultures, which disintegrated in their turn? Why did Persepolis disintegrate? What happened to the ancient civilization of Fatehpur Sikri in northern India? Why did the Hohokam culture of the Arizona desert vanish? What put an end to the empires of ancient Ghana and Zimbabwe? In many cases, of course, the answer is war — and a more powerful adversary. But as Marc Reisner puts it in *Cadillac Desert*, "Explaining the collapse of ancient civilizations is a cottage industry within the anthropological and archeological professions, like the riddle of the dinosaurs." Reisner himself favors a simple answer: increased salinity brought on by poorly drained irrigation.

The real calculus of decadence is almost certainly more complicated than that. But anthropogenic degradation of the environment — disruption of the essential hydrological cycle — must have played some role in the disappearance of ancient cultures. Catastrophic droughts no doubt contributed, aggravated by desertification. Arid lands can be made desert by human action, and deserts can be extended by careless destruction of the soil; deserts where man has been active longest are more arid, more barren, than "natural" deserts. There is some certainty, for instance, that the great migrations of the Golden Horde across the steppe into Mother Russia were set off not by power rivalries or wealthy empires but by aridity and the thirst for water. Human-caused salinity no doubt has its place among the villains. Nevertheless, cultures can survive anything but an outright shortage of water.

If superstition and engineering have been entangled in ancient cultures, so have superstition and politics. The Talmud says no man may sell water from a public cistern; nevertheless, while it is required that the laws of hospitality be followed, the needs of the townsfolk and their thirsty animals must prevail over the needs of foreigners. Even where outsiders' need is great, the person who controls the water has the right to partake of it first. A man in the desert who comes upon a thirsting stranger is entitled to satisfy himself before he shares his water bottle. Charity is a high virtue, but the preservation of your own life is a higher one.

Joyce Shira Starr, whose book *Covenant over Middle Eastern Waters* rooted diligently among the Jewish, biblical, and Koranic water sayings, found an apt quotation from Maimonides' *Book of Acquisition:* "When people have fields along a river they water them in the order [of their proximity]. But if one of them wants to dam up the flow of the river so that his field may be watered first, and then reopen it, and another wants to water his field first, the stronger prevails. The cistern nearest to a water channel is filled first in the interests of peace."

The Koran says no man may abuse a well. If the owner has a surplus of water, he must provide it to strangers and their cattle, but he has no obligation to provide water for irrigating crops. Desert waters are the source of all real property. Thus power politics: the right of first use, the right of might, the seeds of conflict where powers are uneven and needs great, the water politics of the Middle East and North Africa today, and the genesis of water wars.

In our own century, as water use presses inexorably up against water availability, the lamentable political fact has been recognized that some 47 percent of the earth's area is made up of river basins shared by more than one country — more than 60 percent in Africa and Latin America — and that the only accepted body of international law on the subject (the UN Convention on the Law of the Non-navigational Uses of International Watercourses, adopted in 1997) had still not been implemented as the calendar tipped into the new millennium.

There have been many efforts to codify international water law. The UN's offshoot the Food and Agriculture Organization (FAO) published an index in 1978 that listed more than two thousand treaties and "international instruments" dealing in one way or another with international watercourses and aquifers, some of them dating back to the first or second century. Most of these agreements were bilateral, and dealt with shared boundaries or rivers that flowed from one country into another. Most of the disputes in water history (and almost all the cases discussed in more detail later in this book) are generated by the perceived notion of one state that action by another has caused some often unspecified harm. It has been the task of the codifiers of law to try to cobble together a list of general principles from the many arguments.

There are three prime movers among these codifiers: the Institut de Droit International, the International Law Association (ILA), and the International Law Commission (ILC). The Institut is a non-official organization of self-perpetuating and self-defined "internationally worthy jurists" founded in 1873 which has gained some important acceptance: its conclusions have often been cited in diplomatic negotiations. The ILA, by a curious coincidence also founded in 1873 — it must have been catching — is another professional body, numbering around a thousand, drawn from a wide variety of countries. It, too, has no official backing, but its recommendations on international law are also taken seriously. The final body, the ILC, is a creature of the United Nations; its thirty-four members are elected directly by the General Assembly and work on projects mandated by the UN.

All three of these bodies have developed legal principles on water law, each with slightly different shades of meaning. In 1961 the

Institut adopted what it called the Salzburg Resolution on the Use of International Non-maritime Waters, and eighteen years later followed that with a further declaration, the 1979 Athens Resolution on the Pollution of Rivers and Lakes and International Law. The Salzburg Resolution set up the principle of "equitable utilization," by which it essentially meant that sovereign rights over international watercourses are limited by the "right of use" of other states sharing the same water. Disputes should be "settled on the basis of equity, taking into consideration the respective needs of the States, as well as any other circumstances relevant to a particular case." This approach requires maximum flexibility, much discussion, and the setting up of bodies of experts to foster cooperation.

The ILA's Helsinki Rules on the Uses of the Water of International Rivers also supported the equitable use notion. States are entitled to a "reasonable and equitable share in the beneficial uses of the waters of an international drainage basin." The commonly accepted notion of "first in time, first in right" and the "use it or lose it" principle so enshrined in the American West were called into question. The Helsinki Rules made it clear that an existing use may have to give way to a new use in order to come up with an equitable distribution. There was provision for compensation, but "first come, first served" would no longer apply. This notion hasn't gone down so well in, say, Egypt, which might under these rules have to modify or even abandon its millennia-old irrigation practices if, for instance, Ethiopia should want to build new hydroelectric projects on the Upper Nile. That Egypt would get compensation, unspecified, has been brushed aside by the Egyptians, who flatly refuse to acknowledge any such principle.

The ILC, for its part, held that while equitable utilization was all very well, a state had no right to any use if such use harmed another riparian state: "A watercourse state may not justify a use that causes appreciable harm to another watercourse state on the ground that the use is 'equitable' in the absence of agreement between the states concerned." In Article 7 of the draft Law of the

Non-navigational Uses of International Watercourses, the commission was even more explicit: "Prima facie at least, utilization of an international watercourse is not equitable if it causes other states harm."

Stephen McCaffrey, professor of law at the University of the Pacific's McGeorge Law School in Sacramento, California (from whose essay "Water, Politics, and International Law" this quick survey was drawn), is dubious about this whole approach. It may work when the harmed state is weaker than the other, he says, but "in other cases, the ILC's solution is at least questionable. It would appear to encourage a 'race to the river' and to reward the winner. That is, a state that develops first would thereby establish an entitlement to the amount of water used or to the particular use made of the watercourse, and other riparian states would not be permitted to trench upon that amount or adversely affect that use without the permission of the former state."[6] As McCaffrey notes, the ILC reversed its approach in 1994, and its final draft formed the basis of the UN convention. Thus, equitable utilization has become the underlying principle of whatever international law now exists.

McCaffrey points out somewhat cynically that since geography generally has favored mature development of watercourses by downstream states (Iraq in relation to Turkey, Egypt in relation to Ethiopia and Sudan), the upstream states generally prefer "equitable use," since they can thereby lever some advantage for themselves. Downstream states generally favor an intolerance of "harm," since it allows them, generally, to go on doing what they have been doing anyway.

If international law is, kindly put, fluid and uncodified, national laws occasionally go quite the other way.

The case of the Bear Claw Ranch in Wyoming illustrates how arcane (and fractious) water laws can become. Wyoming, the least populous of the United States, with fewer than half a million people, codified its water law when it joined the Union in 1890 with only 10,000 inhabitants. Its system is based on priority: whoever establishes an irrigation right initially has the first call on water from an aquifer, creek, or river. This "number one right" entitles the holder to 1 cubic foot of water per second for every 70 acres of arable land. Wyoming law forbids taking number two rights off a stream until number one rights are satisfied, and also forbids taking water reserved for, say, number four rights. In practice, irrigators take what they can until someone complains, and there is much patrolling of creeks by irate rights holders and alleged midnight irrigating.

The historic Bear Claw Ranch was run-down and decrepit until it was bought by refugees from Hollywood, John McTiernan and his wife, Donna Dubrow, who restored it and began raising the lean meat hybrid called beefalo. The ranch held the number four rights to Smith Creek, a stream running out of the Big Horn Mountains. Because his neighbor, Sam Scott, wasn't using his number three rights, McTiernan used those too. Then he discovered that Scott was planning to subdivide his land and use his number three water to irrigate a new subdivision of "cocktail ranches" or "ranchettes" — 50-acre spreads too small for real farming. That's when the fight began.

McTiernan filed for what Wyoming calls an "abandonment," which in this case meant that Scott couldn't have his water back because he hadn't been using it. This was the other principle enshrined in Wyoming law: use it or lose it. If you don't use your water for five years, you risk losing your rights. Scott testified that he did irrigate his land, and as proof produced records of when he opened the sluices on his dams. McTiernan's expert witnesses declared them a forgery, and he followed this small victory by buying seven NASA satellite images at $7,500 each to show conclusively that no irrigation had taken place. At the end of 1998 the case was still before the courts. Whatever its disposition, the local feeling about the attempt to force an abandonment was summed up by the widely quoted saying, "It may be legal, but it ain't neighborly."[7]

And in the end, it may be academic. Smith Creek's number one and number two water holders have yet to enter the fray. If they do, McTiernan may find his number three rights entirely moot. There is not enough water for everyone.

In Texas, the Edwards Aquifer surrounding the city of San Antonio is "subject to enough depositions to flood Texas," according to *National Geographic.* The farmers want to irrigate with the water; but San Antonio wants the water for industry and for domestic supply to its 1 million people. To complicate matters, the springs drain into the Guadalupe River and provide water for both agriculture and industry downstream. Several endangered plant and animal species also depend on the springs and may die if they dry up. Passions, therefore, run high. In a memorable quote, Texas farmer Maurice Rimkus declared, "Yes, you can have my water, just like you can have my gun . . . when you pry it out of my dead hands."[8]

Water wars can be grand clashings in the international arena. They can also be fought on a small but ferocious scale, with blizzards of paper as ammunition and cadres of bureaucrats as foot soldiers.

PART II

Remaking the Water World

4

Climate, Weather, and Water

*Are we changing the first, and will changes
to the other two necessarily follow?*

THAT CLIMATE governs weather, that weather dictates water distribution, and that water distribution controls life is obvious enough. One has only to look at the fantail beetles of the Namib Desert to understand the relationship. Climate is linked to the rotation of the earth, to wind patterns, to oceans and ocean currents, and to the amount of solar radiation reaching the earth.

Climates change? The El Niño effect, which brings periodic flooding to otherwise arid regions and can abruptly change food chains and ecological systems, is the best-known popular example, but there are many others of varying time lines. About one-third of the planet, for example, has been covered by ice in fairly recent geological time. New anxiety over the instability of the Antarctic ice sheet has led to closer studies of the region, and these have turned up evidence that the ice sheet did indeed collapse at least once, during a recent (by geological standards) interglacial period less than 2 million years ago. One reason the Antarctic sheet could be a worry is its sheer size. Its complete collapse would increase mean sea levels by somewhere between 3 and 5 meters, with obviously catastrophic effects. Another is its composition. The coastal ice sheets are fed by "rivers" of ice flowing from the interior. If the coastal sheets collapse, these rivers would likely accelerate, on what timetable is unknown, but the change could be measured in centuries rather than millennia.[1]

We are, of course, now in another interglacial period. And the critical question is this: Is global warming being intensified by industrial waste gases such as carbon dioxide? Are we humans changing the climate in such a way as to affect both weather and water, and therefore our own lives and futures?

I trudged north from the center of Timbuktu through the dun-colored streets, passed by the old mud-built university that had made the city a center of Islamic learning when it was founded in 1340, skirted a couple of shabby mosques with their distinctive prickling of structural poles protruding into the air, and eventually came to a great depression in the desert floor, maybe 50 meters deep. It was early in the morning. I had left the Prefecture of Police, where I had collected one of the obligatory passport stamps. The area was peaceful again, though Tuareg bands were still said to be harassing travelers in the northern oases, and the police were watchful and suspicious. A block farther on I found a police van stuck in the sand, its wheels spinning impotently. Crowds passed by, paying it no mind and never offering to help — Fulani women in startling scarlet and gold robes; Tuareg men, veiled and hidden, in their distinctive indigo. The sand was 20 centimeters deep, much thicker in places, and crossing it was like walking across a beach; the houses were nearly half a meter below street level. Soon, I guessed, they would be overtaken and buried as the Sahara continued its relentless march. The previous day I'd had a gloomy conversation with the *chef de mission culturelle* of Timbuktu, whose losing job it was to protect the decaying but historic city from poverty, neglect, rapaciousness, and the Sahara.

"Money," he said, "money's the real problem. Without money, you can do nothing. You cannot sweep back the Sahara with a broom."

UNESCO had declared Timbuktu a World Heritage Site, a part

of the cultural patrimony of the human species and therefore worth protecting. But it had provided no money.

"Are things getting worse?" I asked.

"Oh yes, of course," he said. "In the old days the traders came up the Niger River and stayed here before attempting the crossing to Cairo or Marrakesh. They had gold with them, and ivory, and they bought salt. The town was full of rich things —"

"No," I said, "I meant the desert."

"Oh, of course. Once there were gardens to the south, to the banks of the Niger, when canals brought the river water here. Now, there is only sand."

"The Niger is at least ten kilometers away from here, isn't it?"

"Yes. It used to be much nearer. But the desert is closing in."

The poor people were cutting down the few scraggly trees for firewood. They had no other source of fuel and no options; you can't eat goat meat raw.

In the old days, he said, the Tuareg would never cut down trees. They needed the shade where they grazed their stock, and in the wadis the trees grew unmolested. But life was changing. Even the Tuareg needed money, and the only way they could get money was to come to town. There they settled down, and instead of maintaining their lives as poor but healthy nomads, they became slum dwellers. It is a planetwide problem, and the *chef de mission* knew he had no answers. Once, Timbuktu had been a prosperous city of 100,000 inhabitants. It slowly shrank, until barely 20,000 were left. Now the nomads are swelling it again, but it's no longer prosperous. And all these new citizens need water.

Young boys were playing soccer in the depression in the north end of the city, scampering about in the 43°C heat — about 110°F. Beyond them a group of camels snickered and grumbled. At the far end of this depression a pathway became a stairway that spiraled downward. I went to the edge of this deep hole and looked in. Another 20 meters down, a ramshackle construction of poles held a creaking pulley and a massive rope. This was one of the few public wells in Timbuktu.

Ahmed, who had taken me out into the desert a few days earlier, was with me. He had pointed out the communal ovens on street corners; each family was allotted an hour with the oven to prepare food. It was a way of saving firewood. "You have to share," Ahmed said. "You can't waste anything. A boy I know spilled a bucket of water last week and they beat him for it. Two buckets a day and a family can survive. We must teach care with water."

"Is there still water in this well?"

"Oh yes. But soon the men will have to go down and dig another ten or twenty meters."

"It's sinking?"

"The water comes more slowly now."

We went back to the crest of the depression and I stared northward. Footprints led off across the plain. I knew there were still nomads out there. Ahmed's family had an encampment a few kilometers away, their tents and lean-tos anchored in the lee of a dune, their cattle grazing a hundred kilometers farther on. Beyond that, there was nothing between Timbuktu and the dusty little Moroccan town of M'hamid but fifty days' journey by camel, through sand, sere scrub, stones, and secretive caves in the Ahaggar and Tibesti mountains, in which early man had left memories of better days. And oases, of course. It was the occasional oases that made the desert traffic possible. Oases meant water. Deep water, but precious.

Once, there was plenty of water in the Sahara, which was covered in verdant grasslands. The evidence is everywhere: not just the cave paintings or the occasional hippo fossils, but middens of cattle bones and signs that grazing animals had once roved here in large numbers. So did elephants, giraffes, rhinoceros, gazelles, and ostriches. Crocodiles and fish stirred the waters of lakes and rivers. Cypress, oak, lime, alder, and olive trees once grew on the grassy uplands. Now there is only gravel, boulders, mountains, sand, and salt, dry lakes and wadis.[2] No one really knows when it started to change, or even whether this was the first such change — perhaps ten millennia ago, perhaps less. There is some sketchy archaeological evidence that the desiccation of the Sahara forced the emergence of a settled culture on the Nile, possibly Nubian refugees from an early ecological catastrophe, when the grassland herdsmen were driven away to find what water they could. By this reckoning, the flowering of ancient Egypt was a by-product of climate change. Of course, it is an Egyptologist's truism that climate's effects, the annual flooding of the Nile, sustained the culture that did emerge.

Even in recent memory, the Tuareg watered their goats and cattle at shallow lakes. No more.

Although humans are making deserts worse, they didn't cause them.

At the other end of Africa, I once lay nose to nose with a beetle that looked like a spider, on the lee side of a huge sand dune somewhere north of Swakopmund, on the fringes of Namibia's Skeleton Coast. This desert is supposed to be one of the oldest on the planet. As evidence, scientists point to the survival there of peculiar plants, such as the welwitschia, a primitive tree that is largely subterranean, only two leathery leaves visible above ground. This curious plant very likely existed before the evolution of flowering plants, the group to which most terrestrial vegetation now belongs. There are also animals in the Namib Desert marvelously adapted to life with almost no water and extremely high temperatures, including my little beetle, quivering in the sand, its fanlike tail spread in the air, looking for all the world like a miniature maquette for a choreographer of alien ballets.

Most of the earth's deserts are not as old as the Namib. The majority are thought to be of fairly recent geologic origin, probably dating from the steady cooling and consequent aridification of earthly climates which has been the primary feature of the present era (the Cenozoic Era, in technospeak), starting about 66 million years ago. Desert plants likely evolved even more recently, maybe only 10 million years ago, most of them in what used to be the Tethys Sea, a larger ocean along the Mediterranean-Caspian axis.

Most deserts are in the subtropical zones to either side of the saturated equatorial belt. Their existence there is due to something the climatologists call the Hadley Cell, another example of a closed circle in nature. At the equator, where the solar energy reaching the earth is at its greatest, evaporation is obviously fastest, and the air near the ground is heated quickly. It rises, expands, and cools, and when it cools it condenses, and the rains fall in tropical storms, great drenching sheets of water. The now drier air continues to rise, and at high levels moves away from the equator, pushed outward by the wall of continually rising hot air. Over the subtropics it cools and starts to descend. It has already lost most of its moisture near the equator, and as it descends it is compressed and becomes warmer, its relative humidity declining even further. By the time it gets to the ground it is hot and dry; it then moves back toward the equator to complete the cycle as hot, arid wind.[3]

Desert temperatures are extreme. The highest air temperature so far recorded was in Libya, 58°C (136.4°F). The soil can get even hotter, with 78°C (172°F) having been recorded in the Sahara. Night temperatures can be dramatically colder. The lack of cloud cover allows the heat to radiate rapidly outward at night, and desert nighttime temperatures are routinely below the freezing point.

By definition, rainfall in deserts is meager, ranging from almost zero in Libya and some South American coastal deserts to about 250 millimeters in the marginal lands of southern Africa. At Cochones, Chile, no rain fell at all for forty-five consecutive years between 1919 and 1964. Usually, however, rain falls in deserts for at least a few days each year — typically fifteen to twenty days. When precipitation occurs, it may be very heavy for short periods. For instance, 14 millimeters, about half an inch, fell at Mash'abe Sade, Israel, in only seven minutes on October 5, 1979.

Occasionally, the rain that does fall fails to reach the ground because of the heat and the aridity. The potential evaporation rates in deserts can be extraordinarily high, typically 3,000 millimeters a year or more. Death Valley in Nevada has recorded rates of over 4,000 millimeters. Even where rain does fall it is erratic, and in some years may fail altogether. The reasons are not well understood. Regions where the mean annual rainfall is between 250 and 400 millimeters are called semi-deserts; they, too, are barely arable, and humans can survive only by using them as grazing lands for livestock. Or, of course, through irrigation.

Deserts, while entirely natural, can also be of anthropogenic origin — a consequence of human actions. It's not altogether certain whether the Sahara itself is an early consequence of human action; the evidence is very thin, though it hasn't stopped extravagant speculation. But even if humans had nothing to do with "causing" the Sahara, they certainly contributed to its spread and the degradation of the surrounding land in the Sahel. As vegetation is stripped from the land, the surface dries out and reflects more of the sun's heat. This condition in turn alters the thermal dynamics of the atmosphere in ways that suppress rainfall. Other experts suspect that increased dust or other atmospheric pollutants could be causing changes in the climate.[4] The process is called desertification.

The Israelis, who know as much about the process as any people on earth, define it as "the degradation of natural resources in an ecosystem such as soil, water, vegetation and wildlife. Global climate changes as well as human intervention influence the rate at which desertification occurs or is averted. The dry land subregions in the East Mediterranean are characterized by intense sun radiation and the poor quality of water resources."[5] Desertification means declining water tables, as well as the salination of topsoil and the remaining water. Increased erosion and the extinction of vegetation, however caused, make the climate drier. Even within natural deserts, human activities can magnify the effects. The Sonora Desert of the American Southwest, which for centuries had been in precarious ecological balance, with a meager rainfall keeping a complex ecology of plant and animal life alive, has become almost entirely barren through human causes, and its natural wildlife has all but vanished.

Even development that is beneficial on balance can affect the process. Constructed lakes can increase precipitation in some areas, sometimes a considerable distance away — Lake Kariba on the Zambezi River is perhaps the best-known example. But the increased rainfall in some areas is accompanied by reduced rainfall elsewhere. This doesn't mean that there's a fixed amount of moisture to evaporate or rain to fall. Weather is not a zero-sum game. It just means that local consequences of large-scale environmental engineering are often unforeseen. The draining of Sudan's Sudd Marshes, or at least the creation of the Jonglei Canal which will push the Nile's water more quickly through them, will benefit downstream Egypt by ensuring that a larger proportion of the Nile waters reaches its territory. But a lessening of the evaporation from the marshes risks changing the microclimate that makes Sudan's grain-growing areas possible — another source of potential conflict in a situation already fraught with tension.

Population growth, the engine of all increased resource demands, is at the heart of the desertification problem. More people mean more animals, and more vegetation cut for fuel wood and construction. The additional herds needed by greater numbers of humans trample and compact the soil, reducing the infiltration of what little water exists, causing erosion and soil damage. In the Sahel, humans have learned to use cattle dung as a source of fuel, but even there, what sparse vegetation remains is being hacked back for kindling, a violation that leads to increased erosion and wind damage. I remember standing in Ndjamena, the capital of Chad, in the harmattan, a windblown sandstorm, in which the visibility was less than a dozen meters, and watching the soil of the Sahel blowing overhead, to be deposited sometimes hundreds and even thousands of kilometers from its source, saline and leached of any nutrients it might once have contained. It is the classic vicious circle: the desert creates ever more desperate measures by humans to survive, and those human activities create more desert. Only a break in the population upswing can stop the desert's relentless progress.

Climates can and do change, sometimes very rapidly, even without man's intervention. They have changed in recent geological times: the Sahara became desert at about the same time that the Nile cultures were evolving. They have even changed in historical times, and with astonishing rapidity. One of the clearest examples was documented a thousand years ago in a colorful saga about the Norse settlement of Iceland and its colonization of Greenland: the confidence trick by an Icelandic outlaw called Eirik the Red and his son Leif the Lucky.

Eirik was, from contemporary accounts, something of an Icelandic loose cannon, the Norse equivalent of a Wild West outlaw. He and his father had left Norway around 960 "because of some killings," a story suggesting that Eirik had made it to his longboat just before the posse. As a refugee in Iceland, he seems at first to have prospered. He made a good marriage and settled down to farm in the gentle valleys of Haukadale, in the south of the island. But his bad temper got him into trouble again and there was more violence. With another posse of aggrieved relatives sworn to vengeance on his heels, he set sail for a land far to the west that had never been settled, but had been described a century earlier by a storm-swept sailor called Gunnbjorn Ulfsson, who had painted vivid verbal pictures of mountains of ice, towering seas, and ghastly storms, none of which seems to have deterred Eirik. Sailing west along the sixty-fifth parallel, he caught sight of the massive 2,500-meter ice mountain now called the Ingolsfjeld Glacier. He turned south, rounded what is now Cape Farewell, and explored the much less savage west coast. Liking it well enough, he returned to Iceland for one more skirmish with his enemies and to round up a few colonists, ambitious outsiders like himself. To persuade them to depart for parts unknown, he was careful to counteract

Ulfsson's earlier horrific impressions and prudently called the new country Greenland. A contemporary, under no illusions about Eirik's character, left a record of his reason: "He said that people would be much more tempted to go there if he gave it an attractive name."

In the summer of 985 the colonists duly signed up, and a fleet of twenty-five ships set sail from Iceland to the newly named Greenland. Eirik's con game notwithstanding, Greenland was clearly more temperate than it is now. There was sufficient grassland on the western coastal strip to permit the cultivation of grains and the raising of sheep, goats, cattle, and pigs. Modern excavations of the settlement have shown that it was reasonably prosperous. Eirik's own farm had four barns and room for forty head of cattle. The remains of a stone cathedral 30 meters long have also been excavated. The Icelanders, and therefore the Greenlanders, were resolute Christians.

For the next few centuries the settlements were stable. That the Greenlanders explored the Arctic regions is known from the discovery at almost 73 degrees north latitude of a stone cairn dating from the fourteenth century, bearing the runic inscription "Erling Sighvatsson and Bjarni Thjordarson and Eindridi Jonsson built these cairns." The Norsemen also found "Vinland," though the exact location of this land of flowing streams, majestic trees, and wild grapes is still a matter of acrimonious academic debate.

In the fourteenth century, however, the climate of Greenland began rapidly and terrifyingly to deteriorate. For a while, ships still made their way west along the sixty-fifth parallel, but by the middle of the century they had to skirt the increasing pack ice and make a difficult circuitous route along more southerly latitudes. By the turn of the century the Eskimos, who had migrated northward after the previous ice age, began to be forced southward by the cold, and there were repeated clashes with the Norsemen. A decade into the fifteenth century, the Greenland colony was completely cut off from the rest of the civilized world. The last reference to Greenland was in the Iceland Annals for 1410, when an Icelander returning home left a distressingly laconic account of his sojourn, a last elegy for a hardy and energetic people now defeated by nature. His was the last ship out: winter had abruptly shut down Greenland, and the exhausted remnants of the colony presumably perished.[6]

Climate change not of man's making. But it defeated him nonetheless.

Regional climate change can profoundly alter local hydrological ecosystems. It can affect the distribution of water basins, shift groundwater tables, and cause desertification in some places, flooding in others. Global climate change, obviously, will do the same thing, only on a grander scale.

Is climate change, in the form of global warming, happening now, and if so, what will its effects be on the availability of the planet's most precious resource? Are the record high temperatures we have been experiencing over the past decade or so an indication of anything other than a short-term cycle about whose beginning and end we know little? In other words, is it a problem, and if so, is it self-correcting?

Are there in fact more "extreme weather events," to use the current jargon, than there used to be? Are hurricanes worse and more frequent? Typhoons and cyclones? Are tornadoes more common and more violent? Are ice storms more likely than they were? Floods? Droughts?

Is the earth going to turn into Venus, a hothouse of methane and poisonous gases, or Mars, a sterile and airless place, deadly to life? And if any of this is happening, is it human-induced?

One theory of climate change should generate caution in extrapolation from short-term trends. In the 1930s the Serbian mathematician Milutin Milankovitch suggested that basic variations in the earth's movement could affect global climate. He had detected, he said, three cycles: a 100,000-year cycle of the planet's orbit, a 41,000-year cycle in the tilt of earth's axis, and a 23,000-year cycle in the wobble of the axis.

And to introduce another note of caution and put the debate into perspective, consider that a *Newsweek* cover story in 1975 asserted that "meteorologists disagree about the cause and extent of the cooling trend. . . . But they are almost unanimous in the view that the trend will reduce agricultural productivity for the rest of the century." Compare that to Al Gore's assertion in 1992 that "scientists [have] concluded — almost unanimously — that global warming is real and that the time to act is now." So either the world's scientists, as synchronized as a school of fish, are darting hither and yon in overreacting to the latest ambiguous data, or they aren't nearly as unanimous as the media reports, politicians, and Greens would like us to believe.

Until 1998 there was, in fact, some disagreement about what the earth's temperature really was doing — and temperature, you would think, would be one measurement that could be recorded with a degree of certainty. Now it is conceded that the global average temperature rose by around half a degree centigrade during the 1900s. This is a small enough change, but not necessarily reassuring. Between 1940 and the mid-1970s temperatures appear to have been roughly stable, but there is evidence that, especially in the last decade or so, the change has been accelerating. There is still room for some ambiguity. The evidence for change was deduced from measurements taken on the ground, and there are arguments as to whether the temperature of the atmosphere is rising (early studies said no; subsequent studies claimed to have found an error and that the answer was yes; further studies said that the finding of error might have been an error).[7] The Intergovernmental Panel on Climate Change, a UN-sponsored group of scientists, concluded in 1996 that there was discernible evidence that humans have already begun to alter the earth's climate. The IPCC based its conclusions on three "fingerprint" studies: a clear warming pattern in the mid-troposphere in the Southern Hemisphere; a decrease in variations in daytime temperatures; and the statistical increase in hurricanes, cyclones, tornadoes, and destructive floods in many countries in the previous half-dozen years. Recent studies on the Antarctic ice sheet have made pessimists of many scientists who had remained skeptical.

One of the key factors is the rise in TMINs (temperature minimums). Since 1950, night and winter minimum temperatures have been rising, while maximum temperatures are also rising, if marginally less quickly. The cycle is self-sustaining: warming intensifies the hydrologic cycle. A warmer atmosphere holds more water (6 percent for every degree Celsius), and the increasing cloudiness reduces daytime warming and retards nighttime cooling by blocking outgoing long-wave radiation. The increased evapotranspiration and cloudiness also help explain why there is now more heavy rain and flash flooding than there used to be.

Ocean warming has contributed to the drift northward since the 1930s of semitropical marine plants and animals along the California coast. More recently, deep ocean warming has been reported in the Atlantic, Pacific, and Indian oceans, not only in subtropical regions but also near the poles. There have also been radical reductions in the number of zooplankton near the ocean surface, and studies have found sharp declines in mid-ocean nutrient levels (leading one scientist to speculate that the ocean was becoming a "dust bowl" — a peculiar metaphor under the circumstances, but revealing). The Scripps Institute of Oceanography in San Diego reported in June 1999 that many creatures in the deep oceans, particularly those on the seabed, are suffering food shortages. The culprit, it was speculated, was warming surface temperatures, which led to a reduction in the "rainfall" of organic matter that falls to the ocean floor. Speculative data indicate that warming sea temperatures may be the global capacitor (thermal sink) that is absorbing the past century's global warming by "storing" the surplus carbon dioxide.[8] If this is true, and the "sink" stops absorbing and begins transfusing, the earth's climate could change much more rapidly than earlier predicted.

There is none of that famous "unanimity" on the issue, but a consensus is building that the temperature increases resulting from a doubling of atmospheric carbon dioxide will change the world's basic hydrologic cycle by increasing both evaporation and precipitation. Depending on which study you look at, the rise is predicted to be somewhere between 7 and 15 percent. Three research papers published in the journal *Nature* in June 1999 asserted confidently that the "extraordinary levels" of carbon dioxide in the atmosphere were affecting weather, causing higher winds and warmer, wetter winters in much of western Europe and North America. The levels are higher, one study said, "than [at] any time in the past 420,000 years." Rainfall will get heavier, though not everywhere. Rainfall patterns are predicted to shift, though the studies lack precision. Some areas will become wetter, others drier, with profound implications for national water policies. Some water-stressed countries will likely go critical, while temperate regions will experience more floods and "heavy rain events" (which have already increased in the United States by 20 percent since 1900). As temperatures rise, disease-carrying insects such as malarial mosquitoes will become more widespread. Lakes that are now pleasant recreational areas might become disease vectors for bacteria. Already, algae blooms are increasing at an alarming rate.[9]

At a meeting in Japan in 1997, the assembled nations agreed to a treaty that would at least slow the onset of global warming. The Kyoto Protocol would require dozens of nations to cut back on greenhouse emissions, some of them drastically. A few nations will meet their targets, but most will not. Even if they all did, the treaty's effects would be modest; an estimate by Tom Wigley of the National Center for Atmospheric Research suggested that the treaty, fully implemented, would keep the global temperature only four-tenths of a degree centigrade lower than if there were no treaty at all.

Many geographers now argue that prevention isn't enough, and that nations should prepare themselves to adapt to the inevitable changes. Proposed measures include breeding food crops that can better withstand drought, discouraging people from living on flood plains, and making irrigation and water use more efficient. In 1998 Joel Smith, who billed himself as a "climate impact expert" at a company called Status Consulting, pointed out that action now will be cheaper than action later. The bridge between Canada's Prince Edward Island and the mainland is a case in point. It was built on pilings 1 meter higher than originally planned, to ensure that ships could pass underneath it if sea levels rise.[10] So persuasive has the evidence become that a number of major American corporations, including the water giant Enron and the chemical firm Dupont, say that it warrants action immediately, and have announced that they will press Congress to guarantee financial benefits for companies that act, global agreement or not. A few major oil companies, among them BP and Shell, pledged to cut their own emissions by 10 percent from their 1990 levels by 2002.[11]

Curiously, studies show that some places will become much colder as the earth warms up. If melting Arctic ice, for example, interferes with the Gulf Stream and the flow patterns of the major North Atlantic ocean currents, New England, Nova Scotia, Newfoundland, Greenland, Iceland, and much of western Europe will become dramatically colder.

Or they may get warmer. Predictably, skeptics exist. Among them is Eugene Stakhiv of the U.S. Army Corps of Engineers, the army offshoot that has built dams and "groomed" rivers across the United States for more than a century. There are, Stakhiv says, three legitimate ways of looking at climate change: it doesn't matter, or we should wait until we have more evidence, or we need to take precautionary measures. "It's fair to say the Corps is in a wait-and-see mode," he says.

I caught up with him at a coffee break at the same UNESCO conference where I had met Peter Gleick. I had found Stakhiv's own paper a model of clarity and logic, and wanted to tell him so, since neither logic nor clarity was necessarily a hallmark of the papers thus far presented, Gleick's excepted. I asked him what he thought was happening to the global climate.

"Have you read all the studies?" he demanded, peering at me owlishly. Before I could confess that, no, I hadn't read them all, he patted the air at around waist height, about a meter off the ground. "Pile them on each other, you'd have this much paper," he said. "I've read 'em all, and I can tell you, many are useless, some are useful, the data are ambiguous, and some of them are simply outrageous." Since I'd identified myself as a Canadian, he bored in on what he perceived to be Canada's shortcomings. "The Canadian GCM [General Circulation Model] on climate change is the worst of the lot," he said happily, grinning at me. "Hopeless. Apocalyptic. Which is why so many people like it. But we don't agree with it at all. For instance, the study forecasts the emptying of the Great Lakes through global warming. Our most pessimistic estimate, on the other hand, is that there might be a drop of a meter at most. They might even become fuller."

In the paper he delivered to the conference, he said that there is sufficient evidence from recent climate change impact studies "to offer some degree of optimism that well-managed resources systems can withstand all but the most severe climate change scenarios postulated by the leading GCMs. In other words, even as the number of countries and regions that are susceptible to water scarcity and variability climbs, there is a reasonable degree of confidence that vulnerability (the degree of harm or damage) can, at a minimum, be stabilized at current levels, or reduced in most cases."

Several Corps studies on specific river systems in the United States were consistent with a larger study of the Nile, he said. The Nile study showed that even doubling carbon dioxide levels would have manageable effects on the river and only a marginal effect on the Egyptian economy (perhaps a change in the GNP of minus 1.6 percent). The Corps study of the Missouri River system showed it to have the most negative changes of all U.S. rivers, but even there the data were inconclusive and contradictory, ranging from a projected decrease in the Missouri flow of 35 percent to a projected increase of 2 percent. "Water management, which is what we do on a daily basis," said Stakhiv, "deals inherently with uncertainties in supply and demand and with managing rapid change; the key is therefore institutional reform (making managers better able to respond) rather than engineering solutions to problems that might not ever exist."

Later, at a second coffee break, Stakhiv added another point. "Technological advances can mitigate problems," he said. "This is something all the studies ignore. But if you're willing to project climate change into the future, why not also be willing to project technological changes and inventions? And look at the data. Everyone is crying that a crisis is imminent. But it seems that crises are always imminent. They never actually seem to get here. We always seem to cope." He took comfort in the fact that "many of the water use projections in the United States have been, to put it kindly, simply wrong. In fact, water withdrawals are actually declining," he said, citing metropolitan Boston, where proper pricing mechanisms had sharply reduced demand.

Still, Stakhiv conceded, even without climate change, "we will have to find more water. There is still some reason for optimism, but the task will be extraordinarily difficult. Despite the concerted efforts of international organizations to upgrade the water management capacity of the fifty or so developing nations in Africa and Asia, not all will succeed equally. And there is no way of determining in advance where the shortcomings will occur."[12]

A 1998 study by Robert Mendelsohn and his colleagues at Yale goes some way to supporting Stakhiv's skepticism. Mendelsohn is an experienced environmental economist, but his study has been controversial, partly because of its results and partly because it was financed by the research wing of America's power-producing lobby. His conclusions suggest that while some countries and regions may suffer from global warming, the overall economic impact is likely to be zero, a notion that kicked off considerable debate in climatological circles. For example, David Glover, director of the Economy and Environment Program for South East Asia (EEPSEA), carefully refrained from endorsing the study, but suggested that it deserves serious examination: "If its findings are correct, then drastic measures to reduce the use of fossil fuels might not be in the interests of developing countries. Such measures could excessively restrict the energy consumption of poor people and divert resources from more immediate environmental problems like water supply and sanitation."[13]

5

Unnatural Selection

Contamination, degradation, pollution,
and other human gifts to the hydrosphere

W E DRIFTED DOWN the Volga River in a small boat, its
engine throaty but rugged, chugging past Yaroslavl on
our way to Nizhny Novgorod and eventually to the
Caspian Sea, still several thousand kilometers to the south. I had
wanted to travel the full length of the Volga, for in Russia this
isn't just another river. It's the longest river in Europe, true, but
that's not what binds it to Russian hearts. In Russian folklore, the
Volga is mother and mistress, companion and teller of tall tales.
Matushka Volga, Little Mother Volga, is Russia itself, as the Nile
is Egypt.

But past the placidly bucolic village of Plyoss, so beloved of Russian painters, the landscape along the river began to change. First,
the trees disappeared. Rather, they were still there, but they were
all dead. Along some stretches nothing larger than scrub seemed
to survive. The guidebooks gave a clue, pointing proudly to the paper and glue-making factories, phosphorus and fertilizer factories,
and shoe factories that dotted the banks, set down in their twenti-
eth-century ugliness among the monuments to the fourteenth-
century battles against the Tartars at Mogiltzi, the ancient monas-
tery at Navalok, and various estates, monuments, and villages.

On the right bank we stopped at the small city of Gorodets,
where the great Russian hero Alexander Nevsky died in 1263 after a

peace mission to the Tartar capital of Bolgari. Gorodets was first mentioned in the Russian Chronicles in 1152, the year it was founded by Yuri Dolgoruski, but the modern city is a disappointment. There's nothing left of its long history except an earthen mound that is said to be the remains of the ramparts burned down by the Tartars. The factory workers there were on strike, again. They always seemed to be on strike, sometimes because they hadn't been paid for months, sometimes just because the working conditions had become intolerable. At the riverside, a worker handed out pamphlets demanding a cleanup of the environment. "Can you smell the air?" he demanded. "We have to breathe this every day of our lives, and it's killing us. The very air we breathe is killing us."

It was true, the air was thick, with an acrid smell and a chemical taste. On the left bank, a huge chimney was pouring a plume of smoke into the atmosphere. By now, the sight of the thousands of dead trees along the banks had become oppressive. Our boat had traveled for several hours through an eerie stillness — no foliage, no birds, no life. My fellow travelers were depressed. The image the landscape conjured, of a Mother Russia poisoned at the hands of the Russians themselves, was grim.

All the towns along this stretch of the Volga were horrors. Pravdinsk, for example. A 1926 guidebook said that the city had been famous for its lace. Now it seemed to consist largely of a pulp and paper plant and a cellulose factory, until recently named, appropriately I thought, after the dreadful Felix Dzerzhinsky, who ran the Cheka, the forerunner of the KGB, for Lenin. The face Pravdinsk showed to the river was profoundly discouraging: a railway line with a tottering trestle bridge led to the pulp factory; lumber, spilled off the trains, littered the ground and rotted where it fell; and trunks that had been delivered to their destination lay in untidy piles. The factory itself was falling down; elevated pipes, strung between buildings, were sagging, and others had been patched with old tires or were rusting through.

The town of Balakhna, on the right bank about 30 kilometers

from Nizhny Novgorod, was even worse. A mining and smelting center, it was one of the first large consumers of Volga hydroelectricity. But its factories were almost a parody of Victorian industrial exploitation — not only grimy infernos but also falling to bits. Grass was growing from the masonry of the walls, storage tanks were crumbling, roofs were sagging, and smoke poured from windows on all sides as well as from the chimneys. Downwind the air was yellow, a cloud of pollution that rolled over everything in its path, as relentless as a line of Nazi tanks, killing all vegetation in the way — shrubs, bushes, grasses, and trees, all dying. Upwind, scrawny trees still survived. As we passed one factory I noticed a gully leading to the river, a runoff into the Volga; it was steaming, a sulfurous yellow in color. A man was fishing just a few meters downstream. I hoped for his family's sake he would catch nothing.

Many of the ancient chronicles detail saline and "bitter waters"; sometimes they were listed for public edification, so they could be avoided — a kind of early Surgeon General's health warning; in other instances they were regarded as just punishment for transgressions against the gods. One of the signs of the imminent Apocalypse is "the bitterness of all waters," and anyone traveling through eastern Europe, the former Soviet Union and its satellites — everywhere that the command economy model operated, with its callous disregard for anything but narrow-focused abstract principle — could be forgiven for thinking that the Apocalypse was no longer imminent but in full cry. There's hardly a river, stream, or brook that isn't contaminated with the runoff from human misuse, whether industrial effluents, agricultural pesticides and herbicides, or worse. (The "worse" could be bacterial contamination — the river as disease vector — or the dumping of radioactive wastes.) A study commissioned in 1997 by the Czech government found that three-quarters of all surface water in the country was "severely polluted," and almost one-third was so badly contaminated that it couldn't support any fish at all.

Globally, the news is dismal. Ismail Serageldin's World Commis-

sion on Water reported late in 1999 that the Amazon and the
Congo are the only major river systems in the world that remain
healthy. More than half of the world's rivers are now so polluted
that they pose serious health risks. "Some 25 million people are environmental refugees as a direct result of river pollution," the report declared.

Things are better in western Europe, where regulation has been
tougher and the Green movement arose earlier and angrier. Yet,
until the 1980s, a single potash mine in Alsace dumped 15,000 tons
of sodium chloride into the Rhine every day for fifty years, the
equivalent of the combined "natural" loads of the Congo River
and the Mississippi. Not surprisingly, fish species native to the
Rhine are either dead or endangered, although a diligent and encouraging cleanup is under way there. The little Espierre River in
northern France was the dumping conduit for a fertilizer plant
until 1983, when its potassium levels were found to be more than
ten thousand times the naturally occurring levels — about the
same amount as in the entire North Sea.[1] The Danube is an open
sewer for the city of Bratislava, among other places. Coastal pollution is regarded as "severe" in the Baltic, in most of the North Sea,
and all along the Mediterranean littoral, from Spain to Italy. In the
last years of the century, a species of "killer algae" was spreading
through the Mediterranean at about 4 hectares a day and by the
end of 1998 had infected more than 4,000 hectares. The rapid
growth of *Caulerpa taxifolia,* accidentally spread by man, was described by marine scientists as a "major biological invasion," and
was found to be threatening marine fauna between Toulon in
France and Genoa in Italy. Sea urchins were found to be trying to
eat plastic and their own waste instead of the indigestible algae.[2]

In North America, the Great Lakes are cleaner than they used
to be — there is fishing in Lake Erie again — but there are still reservoirs of agricultural and chemical runoff, and heavy metals
in health-hazardous concentrations. Among the worst-polluted
places on the continent are Lake Michigan, southern Lake Huron,
southern Georgian Bay, Lake Erie, Lake Ontario, the lower St.

Lawrence River, the northern coast of Maine, the Atlantic coast from Georgia to Boston, Puget Sound, San Francisco Bay, and the Pacific coastline from Los Angeles to Ensenada, Mexico. Dozens of cities are still pouring raw sewage into waterways. Sewage outfall pipes still dump storm water mixed with sewage into New York's Hudson River, despite the city fathers' loony plan to build swimming beaches in lower Manhattan. Both Halifax, Nova Scotia, and Victoria, British Columbia, Canadian provincial capitals on opposite coasts, have no sewage treatment at all.

In South America, the worst polluter on the continent is Brazil. The cowboy capitalism of the Brazilian interior, and a robust Brazilian disinclination to follow the rules laid down by corrupt officials, have led to an outpouring of chemical and industrial pollution exceeded only in eastern Europe and parts of China. More than 130 tons of mercury are still washed onto the banks of the Tapajoz River every year from the gold mining industry. It is probably no coincidence that a 1994 study of the region revealed an epidemic of birth disorders as well as severe bouts of chemical poisoning among adults.[3]

The worst pollution in Africa is found along the Gulf of Suez and the mouth of the Nile, near Maputo in Mozambique, and around Tunis, but there is troublesome pollution also at the mouth of the Congo River, along most of the coast of South Africa, all of Tanzania north to Ungama Bay in Kenya, at Djibouti, in the Gulf of Aqaba and the Gulf of Suez, and, as we have seen, along the Mediterranean coast.

In China, 80 percent of the country's 50,000 kilometers of major rivers are so degraded that they no longer support any fish. Seventy percent of China's catch once came from the Yangtze, but it has declined by more than half since the 1960s. The rapid industrialization has made the idea of pollution control moot. Along the Yellow River, discharge from paper mills, tanneries, oil refineries, and chemical plants has poured into the water, which is now laced with heavy metals and other toxins that make it unfit even for irrigation. Traces of lead, chromium, and cadmium have been found in vegetables sold in city markets, and so have concentrations of arsenic. Farm chemicals washing into the sea are being blamed for massive blooms of algae. Shanghai has spent some $300 million moving its intake of water farther away from the city because nearby river water was too polluted. The biggest cause of water pollution, however, is sewage. Many of China's rivers contain ten times as many bacteria from human waste as waterways in Western countries. A World Bank study in 1997 put the cost of air and water pollution in China at $54 billion a year, equivalent to an astonishing 8 percent of the country's gross domestic product.

The litany of problems is relentless. In the Mekong River, inland fisheries have dropped by half in the last twenty-five years. In the Ganges and Brahmaputra systems, industrial runoff has increased dramatically and the fecal coliform counts have reached crisis levels. In Pakistan, an emergency was declared in Peshawar when one thousand people were admitted to hospital after drinking poisoned water caused by leaking pipelines. The Yamuna River, which passes through Delhi, receives nearly 200 million liters of untreated sewage every day. Buenos Aires treats only 2 percent of its sewage. The bucolic little province of Prince Edward Island, Canada's smallest, is seeing its gentle meadows and pastures converted by agribusinesses into massive industrial farms, with large areas poisoned by agricultural runoff from massive potato fields — to the ludicrous extent that at fertilizer time local residents have taken to keeping their kids indoors and taping their windows shut to keep the pesticides and herbicides out. The Red Sea is polluted, as is the whole coast of Israel. In Africa, the Senegal and Niger rivers, among others, have seen cataclysmic drops in fishing catches. In South America, Colombia's Magdalena River has seen its fisheries decline by two-thirds in fifteen years because of pollution from oil production. In North America, the Illinois River supported two thousand commercial fishermen a hundred years ago, but the landings have dropped by 98 percent, mostly owing to heavy sewage loads.[4] In Sydney, Australia, tap water was tainted early in 1998 with a dangerous parasite that may have come from dead dogs or foxes, leading to panic buying of bottled water. A Dublin conference on water and the environment in 1992 finished with the depressing conclusion that "after a generation or more of excessive water use and reckless discharge of municipal and industrial wastes, the situation of the world's major rivers is appalling and getting worse."

It is not hard to see why. For centuries, rivers have been used as depositories for human wastes — as natural, God-given sewers, out-of-sight, out-of-mind flushing mechanisms. If you have something you don't want, why, the river will take it away for you. Until recently, this solution caused few problems, and the system maintained itself in balance. Rivers and lakes are naturally efficient at self-cleansing. Natural processes such as sedimentation, aeration, mixing, dilution, and bacterial processing all helped break down wastes and return them to the natural environment. But growing populations and accelerating industrialization have overwhelmed the natural recycling properties of waterways, resulting in gross pollution and increasing threats both to human health and to the water supply generally.

In the west of England, in one of the verdant valleys of that most gentle of landscapes, there is a country house with a wood behind it — not a wild wood anymore, because the old oaks and beeches have long since been transmuted into joists and beams and posts, but not groomed either. It's a mixed forest of hundred-year-old growth — an unruly jumble of larches and imported cedars, of impudent young oaks and elms, of wild cherries and junk woods of no particular lineage. Through the center of the wood, at least in the spring and early summer, runs a brook. The marsh on the other side of the village where the stream once rose has been drained and plowed, and the stream dries up in high summer.

A grandfather was walking in this wood with his grandchildren, a boy and a girl. The children were skipping along the stream banks looking for water spiders, tadpoles, frogs, minnows, anything that lived. But there were no frogs anymore. They had all gone. The grandfather had read somewhere that frogs were dying all over the world. Their damp, smooth skin was apparently too porous for their own good, rather like a lung worn on the outside. The diminishing ozone and increased ultraviolet radiation burned them, and they were transparent to poisons. Nature's early warning system, the grandfather thought, like canaries in a mine — except that they are measuring poisons instead.

The little girl kneeled and dipped her face into the cooling waters.

"Don't do that!" the grandfather shouted, alarmed.

"Why not?" she asked, startled.

"You don't know what's in the water," he said, lamely.

"But I'm thirsty!"

"I know. But you can't drink that. It might make you ill."

"How can it do that?"

"It could have almost anything in it," he said again. "You can't be sure. An animal might have died upstream. Someone might have put something in it that would harm you . . ."

"Why would he do that?" asked the little girl, making an early gender assessment.

The grandfather didn't answer. Why? Because that's what people do. Once, when he was a boy, you could walk through the woods and everything would be fresh and new, the streams clear, full of life. Now . . . Last week he'd found a magpie, its beak held tight by a ring tab from a pop can, drowned in the stream. Even here there was junk: plastic wrappers, cigarette cellophane, a discarded boot. He'd been in the Lake District the previous year, and there the windswept farms and gorse meadows still smelled clean and fresh, as pristine as they'd been made, but he knew it was artificial, a holdout, a redoubt against the tide of the present, held at bay only by bureaucratic fiat, and thus vulnerable.

It was the same all over Europe. No one drank wild water anymore. You never knew what evil was being perpetrated upstream, who was dumping what, who was hiding what by throwing it into the water. He remembered his wife had told him that the last time she had strolled along the banks of the Liffey outside Dublin she had found discarded condoms, pop cans, a rotted plastic sheet, and had been afraid of needles — that she might tread or sit on one and be infected with hepatitis C or AIDS. This was where she had played as a child and swum in the river. Everything was dirty now. The grandfather supposed it was inevitable.

It was the same everywhere. He had visited his daughter in Toronto the previous year, and they had gone down to the beach on Lake Ontario, but there was a sign that read: "Beach Closed. Unfit for Swimming." His daughter had shrugged. Toronto was downstream from Chicago, Detroit, and the Love Canal, though she had heard Chicago had cleaned up its lakefront by dumping its wastes into the Mississippi headwaters instead, letting St. Louis and New Orleans inherit the worry. A report of the International Joint Commission late in 1998 had said that the Great Lakes were "cleaner than they had been for fifty years," but that wasn't saying much. Sure, you could no longer set a match to Cleveland's Cuyahoga River and watch it burn, and the piles of dead fish on the shores have been cleaned up. The Love Canal scare had receded. But there was still toxic sediment in dangerous concentrations in all the lakes, and the beaches were often closed. Excessive bacteria levels caused by Toronto's own sewage and storm drain runoff were blamed.

A report by the U.S. National Resources Defense Council said that there had been 4,153 beach closings in the United States in 1997, more than double the number of the previous year. The increase had apparently been caused by higher rainfalls, but also because more states were monitoring their water and finding it was sub-par. None of this surprised or particularly alarmed anyone. It was just the way things were. You can always swim in a pool. The chlorine levels there would kill almost anything. Sometimes, though, she wondered, where did the chlorine go when they drained all these pools? A special report on tap water safety in the United States found lax standards and careless monitoring of the nation's drinking water: about 40,000 of the 170,000 water systems in some way violated purity levels. Of these, 9,500 water systems serving almost 25 million people had "significant" violations, defined by the Environmental Protection Agency as posing "serious threats to human health."[5]

Pamphlets handed out to Ontario visitors hoping to try their hand at sport fishing make sobering reading. "Ontario: Yours to Discover" is the provincial slogan, and one of the things to be discovered is that the province's fish should be eaten with care and in small quantities. Not because the fish are in danger of extinction, but because eating too many of them can do injury to human health. Methyl mercury, among other poisons, has been accumulating in the biomass, and several northern communities have been found to exceed the World Health Organization's recommended standards for mercury.

The grandfather from England who couldn't swim in the Great Lakes of North America was not alone. There are more than 8,000 kilometers of shoreline on the U.S. side of the lakes, but only 3 percent is fit for swimming, for supplying drinking water, or for supporting aquatic life. In 1993, two-thirds of U.S. "fish advisories" were issued in the Great Lakes region, most of them having to do with excessive amounts of mercury, PCBs, chlordane, dioxins, and DDT. Each year, 50 to 100 million tons of hazardous waste is generated in the watershed for the lakes, 25 million tons of pesticides alone. There are airborne pollutants, too: trace pollution has been detected from farms in Mexico and cement factories in Texas.

The 1998 International Joint Commission, in the same report that declared the Great Lakes cleaner than ever before, tracked the accumulations of a dozen or so noxious chemicals and found, to its dismay, a "buildup of radioactive contaminants in the lakes from nuclear power plant discharges. The evidence is overwhelming," the report said. "Certain persistent toxic substances impair human intellectual capacity, change behavior, damage the immune system, and compromise reproductive capacity." Anyone relying on fish or wildlife for food was at risk.

By mid-1999, Ontario companies had violated the pollution control regulations 2,000 times, more than twice as many violations as in the entire previous year.

Elsewhere, industrial pollution by lead, mercury, and cadmium are becoming ever more serious problems. Cadmium causes renal disease, and cadmium deposits in the Rhine increased twentyfold between 1920 and 1970, from less than 2 milligrams per kilogram to more than 40. Mining is causing extensive mercury poisoning in hitherto pristine Amazon waterways. Lead is prevalent in many rivers in industrial countries, causing kidney trouble and damage to the nervous system. Aluminum, which earlier had been thought harmless, is now believed to contribute to Alzheimer's disease, and is increasingly common as industrial waste.

There is no "them" doing all this. We're all at fault. It's the nature of our civilization — a point Canadian businessman and environmentalist Maurice Strong has been making for years in a series of speeches. Environmental degradation is not a "problem" to be "fixed"; it is a sign of a fundamental imbalance in the way our industrial civilization is run. I've been a part of the process, so I know how it works. When I was living in Toronto in the 1970s, we found termites in our house. This was an unnerving feeling, because termite damage is almost always discovered when something collapses; termites' need for cellulose and for constant moisture causes them to eschew open air, sending them burrowing into wooden members from the inside, where they can't be seen. The first a bewildered homeowner knows of their presence is when floor joists give way, or a nail driven into a bearing member goes right through because the wood is hollow. In our case, the wooden sills holding up one corner of the house were so rotten I could punch through them with my fist.

The termites infesting Toronto were more difficult to eradicate than the tropical species I was familiar with from growing up in Africa. There, at least, they built their mounds on the surface and could be eliminated if necessary. In Africa, we mostly left them alone, knowing they performed a useful ecological function. Only if they invaded a house were steps taken — or if someone wanted a tennis court; termite mounds, crushed and rolled, made the best clay tennis courts possible. But the Toronto termites were much harder to deal with. The queen, the nerve center of termite life, could be 30 meters or more underground and as many meters laterally. Queens were virtually impossible to kill. The only solution was chlordane, one of the most potent pesticides ever made, and one of the most persistent. We used chlordane precisely because it never deteriorates.

The process was to drill a ring of holes around the perimeter of the house both inside and out, to a depth below the footings of perhaps 2 meters. Then several dozen liters of chlordane were pumped in at high pressure, in effect forming a "skin" of chlordane around the building. We knew what we were doing — poisoning the earth. We knew enough that we never ate tomatoes from our garden again. But the choice seemed stark: poison a patch of the earth or lose our house.

It seemed like an easy decision at the time. We couldn't afford to lose our house. It would be the easy decision for a nation, too, or a culture. Haven't you heard the political speeches: we can't afford to lose our industries, our jobs? It's the same choice the human species has been making since pesticides were invented.

Of course, it is entirely the wrong decision. Eventually the chlordane thins out. It migrates downward, then reaches the water table. In the cities no one is using wells, so if the water table is contaminated, it doesn't pose an immediate risk to human health. But as the termite infestation spreads, so does the use of chlordane. Eventually it makes its way, say, into Lake Ontario, where it joins the accumulating poisons from other jurisdictions. Toronto gets its drinking water from Lake Ontario. The solution is to push a pipe farther and deeper into the lake, where the water is still clean — sort of.

It's a solution followed elsewhere. When the city of Halifax hosted a meeting of the G7 leaders, the town fathers didn't want raw sewage lapping the shoreline where the leaders would meet. The solution was not to treat the sewage — that would have been too expensive, and they didn't have time. So they extended the outfall pipe farther out to sea.

Bio-accumulation and bio-magnification make everything worse. Chemicals become more concentrated as they pass up the food chain. You would have to drink water from the Great Lakes for a thousand years to take in the same amount of PCBs you'd get from one fish a kilo in weight.

Dozens of epidemiological studies have confirmed links between cancer or birth defects and contaminated groundwater. The *American Journal of Epidemiology* in 1983 conducted a study in San José, California, and found a statistically significant increase in birth defects owing to solvent-contaminated groundwater.[6]

In developing countries things are generally worse, and as development accelerates, the amount of water contaminated with industrial pollutants such as petroleum and toxic metals rises sharply. Perhaps the overall national water supply is no dirtier, but there's very little clean water, and the meticulous scrubbing that water gets in the cities of the developed world simply doesn't exist. Mostafa Tolba, the former director general of the United Nations Environment Program and a grand old man of the environmental movement, admits that despite all efforts, the quality of fresh water depends not on pristine supply but on decontamination. "The number of polluted rivers is on the increase," he warned a 1998 UNESCO conference. Peter Gleick was equally blunt: "According to the latest data, most of Africa, most of Asia and most of western South America fail the simple test of providing water for basic drinking and sanitation needs. Over a billion people have no access to clean drinking water, and more than 2.9 billion have no access to sanitation services. And the sanitation trend is getting sharply worse, to some degree only because of better and more accurate reporting, but also because of higher populations, especially in the cities."

If you drive north from the center of the Cameroon city of Bafoussam, heading for the highlands and Lake Chad beyond, you'll pass an apparently endless stretch of ghastly slums, over-crowded shantytowns, tumbledown and wretched, hundreds of children playing in the fetid stagnant pools and among piles of rotting garbage. On the main roads, markets are set up alongside refuse dumps. When cars cease running, they are simply aban-doned on the roadside and are immediately colonized by families — a mother, father, and children making do in a space not much bigger than a sofa. The soil is black with discarded crankcase oil, half-burned debris, and general filth. Peasants come in from the countryside and lay their piles of goods to sell in the sludge of the sidewalks — here a small pyramid of mangoes, a startling flash of green and gold among the dun; there a row of fish, still shining but already beginning to turn in the heat; here an old woman wearing bunches of bananas like shackles. Diesel fumes from badly tuned motors fill the air, and most of the buildings are crumbling and black with soot. Off the main roads, if you plunge into the maze of shanties, cobbled together with strips of iron, old wire, cardboard, and flattened tin cans, the stench can be overpowering. There are no services here: no roads, no addresses, no sewage, and no water. Water has to be fetched from communal standpipes, dripping taps in a circle of mud, ankle deep and stinking.

Not to pick on Bafoussam. It is remarkable only by contrast with its beautiful hinterland, rolling hills covered with emerald forests and bright scarlet blossoms of improbable size, and the peasant architecture in the area, which is striking: steeply pyrami-dal roofs, sturdy little houses of beige mud brick, their yards swept clean, an almost Disneyfied cuteness to them. Otherwise, Bafoussam is very like other cities in Cameroon — Douala or the capital, Yaounde. It is also like the shantytowns of Cape Town, the slums of Luanda, and the megaslums of Lagos, Kinshasa, and Dakar. Megaslums also exist all over Asia — in China, Thailand, Laos, India, Bangladesh, and Pakistan — and in almost every

country of Latin America. In other words, they are found throughout the developing world.

Anytime now — no one knows exactly when — the world will
pass a significant human milestone. For the first time, more people
will live in cities than in the countryside. And most of the growth
is happening in the developing world, where cities are exploding at
a rate that has made it impossible for urban services, including
supplies of sanitary water, to keep up. In the early 1990s an apocalyptic literature began to emerge in the West about these burgeoning slums of the Third World. Some authors seemed obsessed with
social decay and degradation, and made odious predictions of
global collapse. I don't buy into this gloomy millenarian view of
the inevitability of disaster, partly because if you talk to the residents of these slums instead of just driving by, it's easy to see that
they're as eager for betterment as any family in the leafy middle-
class suburbs of, say, Connecticut; and partly because the writing
shares so many hand-wringing attitudes with that of the early Vic-
torians in nineteenth-century England who were convinced that
bringing together millions of the brutish poor in the slums of industrial cities was bound to mean the end of civilization as they
knew it.

Which is not to say that the human misery in these slums
should be easily dismissed.

For most of human history, cities have been the repositories of
power and wealth but also of plagues and pestilences. It was little
more than a century ago when the major cities of the industrializ-
ing world introduced sanitary sewers and decent fresh water, and
the incidence of disease began to decline. In 1852 the average age at
death in the boomtown of Dudley in England was seventeen,
mostly because of the "human excrement in all back streets, courts
and other eligible places."[7] In the two decades after sanitary sewers
were introduced, life expectancy increased by a dozen years. In
French cities life expectancy rose from thirty-two to forty-five be-

tween 1850 and 1900. Of course, advances in medical sciences and the understanding of diseases helped; a cholera epidemic in London was famously traced to a single public faucet. But the mere fact of sanitation was the greater cause.

In the rest of the world, most of the population still lived in villages and on small landholdings; high disease rates were generally associated with the poverty-stricken life of the countryside. But the same shift that happened in industrialized Europe and America in the nineteenth century is now happening in the rest of the world, only much faster and on a hugely increased scale. It took London from 1800 to 1910 to multiply its population from 1.1 million to 7.2 million, a growth rate that has been achieved in some African cities in a single generation. Many Asian cities have increased fourfold in the same period. Rapid population growth and rapid urbanization are twin phenomena, the entire process taking place in countries that are poor. Already, twenty-two cities in the world have populations exceeding 10 million, and seventeen of them are in developing countries. The largest city in the world, Mexico City, has more than 20 million inhabitants. The same thing is happening in Delhi, Amman, and Santiago, where the urban poor are already somewhere between 30 and 60 percent of the population. In Addis Ababa the figure is 79 percent; in Luanda, 70 percent; in Calcutta, 67 percent. The population growth rate in the slums is higher than elsewhere; in Bangladesh, for example, it is four times the rest of the country's birthrate. If present trends continue, which seems probable, more than half the inhabitants of these megacities, well over a billion people worldwide, will soon live in slums and shantytowns without any amenities whatever. UN figures show that the world's rural population will peak at 3.1 billion by 2010 and decline thereafter. By 2030, by contrast, the urban population will be twice as large, and cities will have grown a staggering 160 percent in 40 years. This rapid increase has made it almost impossible to keep up with need, despite heroic efforts. The 1980s, the UN's "International Drinking Water Supply and Sanitation Decade," saw an impressive 80 percent more urban dwellers gaining access to municipal water supply and 50 percent more to a system of waste disposal. But by 2000, an extra 900 million people had been added to the planet. The actual number of those without water remained the same, and the number of those without sanitation rose by 70 million.[8]

In many cities the authorities have almost given up. Their water infrastructure is in such bad condition that it cannot possibly recover its costs from users, and their income is far too low to improve the service. The quality therefore declines further. There are disruptions and occasional sabotage by disgruntled consumers, or the system is damaged further by the desperate trying amateur diversions. Often, as in London, nearly half the water simply disappears through old, cracked, and leaking pipes.

Until recently, development planners seldom paid much attention to drainage and water treatment issues. In Latin America, only about 2 percent of human waste is treated. Untreated waste flows into the river systems, forcing cities to go farther afield for their intake; in Lima, pollution has upped treatment costs 30 percent.[9] The same thing is happening everywhere in the developing world.

Ironically, given the falling water tables in so many parts of the world, a growing problem in these new massive slums is the reverse: rising water tables. This is caused partly by the destruction of natural drainage systems such as wadis, but there is another more dire cause. The lack of adequate sewage systems has meant that most residents rely on cesspools, essentially concrete tanks without a bottom, whose stinking (and unhygienic) exfiltration seeps into the groundwater, causing it to swell.

Pollution leads to disease, a straight-line computation that is self-evident. Best guesses are that some 250 million new cases of waterborne diseases occur every year, killing somewhere around 10 million people — as if all of Canada were wiped out every three years.

Cholera is the worst news. The first cholera pandemic spread to the Americas from Asia early in the nineteenth century, and until the American Civil War there were regular cholera epidemics in the United States. In 1970 the disease invaded Africa and in 1991 returned to the Americas. It spiked in 1992 and is still going up. In 1990 there were about 100,000 cases worldwide, but in 1992 the number exploded to 600,000, the overwhelming majority in the burgeoning slums of Latin America.

Bilharzia, or schistosomiasis, is the other major water-borne disease, currently infecting more than 200 million people in some seventy countries. The disease is spreading and intensifying, mostly because of otherwise beneficial projects that give the host snails a comfortable new environment. In the Nile Delta, for example, the snails spread rapidly after the construction of the Aswan High Dam and its irrigation channels, leading in some areas to infection rates of nearly 100 percent. The same thing happened in Sudan after the construction of the Sennâr Dam, and after Lake Volta was constructed in Ghana, where infection rates in children jumped from less than 10 percent to more than 90 percent within a year.[10] Lake Malawi is now said to be the only sizable body of water in Africa that is bilharzia free, and therefore safe for swimming. Tourists gamboling on its beaches would be unwise to bet on it, though.

In the developed world, so-called Legionnaires' disease is also caused by contaminated water. Its incidence is still low but worrying; it is resistant to chlorine, and so can infect water treated to U.S. and European health standards.

In April 1999, *New Scientist* reported that *Helicobacter pylori*, a bacterium involved in a wide variety of gastrointestinal diseases, had been found in a large number of water delivery pipes tested, growing in the slime that often coats pipes and that is resistant to chlorine.

Malaria is also, of course, a water vector disease, carried by the mosquito. More than 2 billion people worldwide are exposed to malaria, and there are about 130 million new cases every year, mostly in Africa and Asia. Malaria accounts for almost one-third of all childhood deaths.

Not all the biological news is bad, however. The other major parasitic scourge, dracunculiasis, or guinea worm disease, is down from nearly 10 million cases a year to near zero.

The changes that humans have wrought are not trivial. The chemistry of groundwater is being modified on a global scale. Human actions have already caused the mean average dissolved "additives" to increase by 10 percent, and in the case of salt, sodium, chlorine, and sulfate by as much as one-third. Leachates from fertilizers, herbicides and pesticides, toxic organic and inorganic chemicals, and radioactive traces are found throughout the world. The situation is worst in the United States, mostly because the chemical industry is at its most inventive there and farming has come to depend heavily on pesticide use, and partly because there is hardly a river left undammed or an aquifer untapped. In the American Southwest, aquifers are being recklessly mined with no hope of recovery. In the Northeast, salt spread on roads in the winter has contaminated drinking water supplies, as studies have found on Long Island. The rapid runoff, or excess water, resulting from deforestation is a problem throughout North America. Irrigation return-flows that leach salts from soils in semiarid areas are major sources of pollution in the western United States, just as they are in western Australia.[11]

All natural watercourses contain some nitrates, albeit in dilute form, that are generally released into the water by storm runoff, since nitrate ions travel freely through the soils. But natural concentrations of around 0.1 milligrams per liter are seldom harmful. That is just as well, because nitrates in the bloodstream can inhibit the ability to carry oxygen ("blue babies" suffer from a condition usually caused by overly high nitrate concentrations in their blood); nitrosamites, nitrogenated compounds, are among the most potent carcinogenic substances known. The nitrate content of fresh water is rising steadily because more and more farmers are using nitrogen-based fertilizers, and levels of industrial and urban wastes are increasing.

Rivers in the richer countries have, by and large, become steadily cleaner over the past decade. But when measured for nitrates, fewer than one in ten European rivers is any longer "natural"; most have nitrate levels four times the norm found in nature. The U.S. Geological Survey found that almost one-quarter of the wells it analyzed in 1990 had nitrate concentrations higher than natural background levels, some dangerously so. In Denmark, nitrate levels have trebled since the 1940s.

On a global scale, rivers are dumping twice the amount of organic matter into the oceans that they did in prehuman times; the flux of nitrogen and phosphorus has more than doubled. The eutrophication, or algae proliferation, of marine systems has become a problem for fisheries worldwide, and estuaries in the developed and the developing world are becoming polluted to a degree that is troublesome to human health. You don't have to agree with Paul *(The Population Bomb)* Ehrlich that economic growth "is the creed of the cancer cell" to know that humans are seriously skewing the very resource on which they are so utterly dependent.

6

The Aral Sea

*An object lesson in the principle
of unforeseen consequences*

EVEN A LIGHT WIND picks up the salts and the dust and turns the air hazy, like heat shimmering in a mirage. When the winds are stronger, the white clouds drift higher and deeper, and the sky is opaque, like milk made of vinyl, impenetrable and ominous. Running north along the Amu Darya River northwest of Bukhara to the Aral Sea — or what used to be the Aral Sea — the farms are also white, dusted with what looks like fine snow. But it's not snow. It's salt, leached to the surface after decades of careless irrigation. Farther on, past the dreary little Uzbek town of Nukus, is Muynak, which was once the shore of the Aral, then the world's fourth-largest lake. The view from the ridge just outside town is perhaps the most famous in the depressing annals of environmentalism: as far as you can see there is nothing but sere sand, white bones of dead cattle poisoned on toxic plants, the rotting hulks of fishing vessels and barges, the pathetic detritus of a once thriving fishing culture. In the foreground are the broken timbers of what used to be a wharf; beyond that are six or seven ships' hulls, grounded at random, pointing no way in particular, some of them still upright, but broken, the paint reduced to small scraps of rust. Farther on are humps in the sand where other hulls have been buried, and beyond that only dunes, sand, salt, and nothingness.

*   *   *

The Aral Basin is the catchment and drainage area of two river basins, the Amu and the Syr, which together contribute most of the water in the Aral Sea. The two rivers rise in the Tian Shan and Pamir mountain ranges of the Himalayas and flow northward through their alluvial valleys and the Kara Kum and Kyzyl-Kum deserts before emptying into the Aral, one into the southern tip and the other into the northern. These two rivers are the only real source of fresh water in a system that comprises 1.5 million square kilometers and somewhere around 35 million people. It used to be bounded almost entirely within the old Soviet Union; the Russians split the region into several ethnically based "autonomous" republics, on the tried-and-true principle of divide and rule. But when the "Evil Empire" split, it left a scattering of tiny, poor "independent" entities — Turkmenistan, Uzbekistan, Kazakhstan, Kirghizstan, and Tajikistan — all of which jealously guarded and maintained their Soviet-created "national" identities. It is these statelets that are left to wrestle with the problems Sovietism left behind. For the Aral is really an ex-sea now, a shrunken thing, and poisonous. The water management disaster that brought it to this state has been described by horrified observers searching for an appropriate metaphor as "the quiet Chernobyl." This may be a little over the top, but not by much, and when you factor in the grim datum that the Soviet military is known to have buried on an Aral island (now a former island) enough anthrax spores to eliminate the entire global population, maybe "over the top" isn't nearly high enough. Without any exaggeration, the Aral Sea has become the greatest man-caused ecological catastrophe our benighted planet has yet seen, an awful warning of the consequences of hubris, greed, and the politics of ignorance. It is also proof positive of how suddenly ecological catastrophes can happen, and how hard they are to reverse once they are under way.

As though proof were still needed.

I flew into Bukhara from Samarkand late one October afternoon on a rattling little Aeroflot flight out of Dushanbe. There was a

dust storm in the region, and a couple of old bandits had taken
refuge in the hut that doubled as a terminal building, their don-
keys tethered outside and their ancient rifles, gleaming and well
oiled, propped against a stack of fading posters left over from Soviet times. The air was thick with roiling clouds of sooty dust
called a "black blizzard" by the locals. Dust choked everything in
the town. There was grit on the floors, grit on the hotel beds, grit
in the cars, grit on the restaurant tables, grit in the food. The sun
was still up, but it was already dark, like a badly lit film noir. I went
to have supper in a place someone had told me about, deep in the
casbah, a sunken well of a place that seemed to have been converted from a cistern. The only other clients were old men drinking tea from glasses and a few soldiers, one of whom got up while I
was eating and stomped a mouse to death on the mosaic floor,
leaving the carcass there, perhaps to be cleaned up the next day, if
there were cleaners around, which wasn't obvious. The waiter had
swept the table more or less free of dust with his sleeve as he
plunked down a plate of lamb shashlik, grilled to stupefaction, and
a Russian beer, which was warm and flat. There was sand in the
glass, too.

I was actually in town on other matters — I was following the
trail of Tamerlane, the greatest of the Golden Emperors of the
Steppe — but the environmental degradation was so obvious it
was impossible to ignore. It also seemed impossible to believe that
this place, which had been desert for longer than human memory,
had been chosen by the central planners of Soviet Russia to be one
of the showpieces of collectivist agriculture. Burying the capitalists
would be tough enough in the rich black loam of the Ukraine, I
thought, but to make this desert bloom required the skills of a nec-
romancer, not an agronomist.

There had always been some agriculture in the region, and
traces of irrigation have been found dating back five thousand
years. Most of the inhabitants had been nomads, but when the
hordes set up their capital in Samarkand, and Tamerlane began
building his magnificent palaces and public buildings, water was

taken from both rivers for food crops. It was sustainable irrigation, the scale appropriate to the resource. The Aral was hardly touched.

When the Bolsheviks overthrew the czars and began to impose their command economy on a bewildered and resentful peasantry in Russia, Central Asia wasn't spared. The planners couldn't believe their luck. All that land not being used! All those primitive peasants not knowing which end is up! All that water flowing uselessly northward, going nowhere! The Gosplan people in Center, as Moscow was called in the hinterlands, got out their maps and their rulers and their abacuses and worked it all out. Products were needed for export. Inexplicably, despite the manifold benefits of collectivization, the Russians couldn't seem to feed themselves, never mind growing things for foreigners. But cotton — there was a crop! White gold!

In 1937 the Soviet Union became a net exporter of cotton, and a decade later mechanized agriculture was a fact of life in Turkmenistan and Uzbekistan. Not just cotton but wheat and other food grains too. There were cattle ranches in this place where nothing grows, and corn fields. It was wasteful, sure; as much as 60 percent of the irrigated water never got anywhere near the crops, some of it disappearing in the unlined canals or through evaporation, and some diverted by bandits just for the hell of it. Nevertheless, the whole scheme seemed to be working. There was lots of water, wasn't there? By the 1950s, Khrushchev could say in all seriousness that he would overtake the capitalists within a decade and mean it.

In 1956, with great fanfare, the Soviet Politburo dispatched its agricultural commissar to open the Kara Kum Canal, which taps into the Amu Darya near the Afghan border and meanders through the desert toward the Turkmen town of Ashgabat, which is not far from the border with Iran. Flying over it a dozen years later was an extraordinary experience. The land was grim, black stone and dunes the color of dirty khaki, but the canal itself was an emerald necklace on a skin of dun, and radiating out from it were the farms, laid out in Soviet fashion in immense and unwieldy blocks, the plowing haphazard and the furrows crooked, as though the farmers had stared too much at the sun.

It was a time of optimism in world agriculture. The Indians were using new seed grains to make their deserts bloom; the Americans were diverting ever-increasing amounts of water through ever-increasing hectares of inhospitable desert; the Israelis were exporting oranges, of all things, and had the romantic vision of turning the Negev into a garden. Anything seemed possible, and though Khrushchev had been forcibly retired to a smallholding outside Moscow, his boastfulness was still Soviet policy. There were rice paddies, after all, in Turkmenistan. The Soviet system could do anything.

Soon afterward three other canals were opened, diverting more water into the Uzbek and Kazakh deserts, irrigating more millions of hectares. But the water that spread across the desert never reached the Aral, and after 1960 the level of the sea began to drop. Before the diversions, some 50 cubic kilometers of water a year had reached the sea; by the early 1980s, the total was zero.

The village of Muynak was an island in the Amu Darya Delta in 1956. By 1962 it was already a peninsula. By 1970 the sea was 10 kilometers away, and the fishing wharves and landing areas were long abandoned. A decade later the nearest water was at a distance of 40 kilometers, and by 1998 it was 75 kilometers. The seabed had become a desert. As the Aral shrank, its salinity increased, from an area of 1,075 cubic kilometers with a salinity of 10 grams per liter to 54 cubic kilometers with more than ten times that salinity.

By 1977 the commercial fish catch had declined by over 75 percent, and a few years later, in 1982, commercial fishing ceased altogether, shutting down an industry that had caught a sustainable 50,000 tons a year and had employed nearly 60,000 people. A few species survive in the three small saline lakes the Aral Sea has become, but most of the fish have died. The fish cannery in Murnak tried to stay in business by buying fish at great cost from Murmansk and Arkhangelsk, the Soviet Arctic ports; but when the Soviet Union collapsed, that lifeline dried up too.

The collapse of the fishery wasn't the only problem, or even the worst. There were dozens of negative secondary effects. The declining sea level lowered the water table in the region, destroying many oases near its shores. Un-irrigated farms dried up. Wildlife disappeared, and the once flourishing muskrat fur industry collapsed too. Soviet agriculture had been pesticide dependent — and, indeed, farms in the area still use DDT, long banned elsewhere. Heavy concentrations of pesticides had built up in the sea. Over-irrigation caused salt concentrations in many agricultural areas.

The Aral had once regulated the climate in the area. It was large enough to act as an air conditioner in summer and as a heat sink in winter, buffering the winter storms from Siberia. With the loss of the Aral, the local climate became more continental, with extremes of temperature, shortening the growing season to 170 frost-free days a year, fewer than the 200 needed for cotton. As a consequence, more farms switched from cotton to rice — which demanded even more diverted water. The exposed seabed, now over 28,000 square kilometers, became a stew of salt, pesticide residues, and toxic chemicals; the strong winds in the region pick up more than 40 million tons of these poisonous sediments every year, and the contaminated dust storms that follow have caused the incidence of respiratory illnesses and cancers to explode. Some of the dust has been reported as far away as Belarus, 2,000 kilometers to the west, and in Pakistan to the east. As the deltas dried up and the wetlands disappeared, the local rainfall decreased by half.

If you stand on what had been the shore of this once great lake, you can see the salts in the air. Throat cancer, the local doctors say, has reached epidemic proportions. So have kidney and liver diseases, arthritis, chronic bronchitis, typhoid, and hepatitis A. It is worst in the Amu Darya Delta, a region called Karakalpakstan, where the bio-accumulation of toxins is so great that mothers have been advised not to breast-feed their babies, and infant mortality has climbed past one hundred deaths per thousand births.

Now that the fish buyers have gone, only the foreign ecologists come, to stare appalled at the ruined landscape, impressed and depressed by the scale of the catastrophe. I find in my notebooks this thought, scribbled after talking to an Israeli hydrologist who had seen what there was to see and was returning with a message of gloom for his colleagues in Israel and America: "Maybe that'll be the Aral's only legacy. It will help make things better elsewhere by frightening people half to death."

Even the thickheaded Soviet planners came to see that something was seriously wrong.

I interviewed the Gosplan folks in Moscow in 1970, and I recently looked up my notes. In the hours I spent with them, they said nothing that was at all negative. Officially, the line was still the Khrushchev one, and the planners dutifully trotted out the statistics: so many tons of cotton, a harvest so many tons bigger than the year before, so many hectares newly planted. You could almost see the smiling peasant girls of Soviet myth, baskets of white gold in their burly forearms, while in the background tractors from the Volgograd factories roared and belched, symbols of industrial might. It was all a fraud. They knew it and I knew it, but in the final days of Brezhnev, that was how it was done. Truth was in the interstices between words, and had to be inferred from silences, shrugs, and things unsaid.

The week before my interview I had read some hints of difficulty in a Soviet newspaper, *Literaturnaya Gazeta,* which had recently strayed from its literary mandate to tackle the tricky subject of pollution. A team of reporters from the *Gazeta* had found pollution staining some of Russia's finest rivers and had traced the origins of one mysterious slick on the Volga to a toothpaste factory, which had dumped its entire year's production into the river when it ran out of tubes.

I shoved the clipping over at the planners. "It says the fish are dying in the Aral," I said. "What's happening?"

They wouldn't say, and fobbed me off with some lame story about natural cycles. But they knew, all right, for down the hall others were already working on a remedy, a "solution" typical of the thinking that could set farm quotas without ever looking at the farmland itself. On the map it seemed obvious. There was lots of water. Siberia had plenty of rivers it wasn't using. Take the 5,000-kilometer Ob, for example, which flows into the Ob Gulf in the Arctic, a chilly and sand-choked inlet unfit for any purpose. The Ob and its tributaries rise near the Kulunda Steppe, to the northeast of the Aral Basin. Why not simply divert it southward? Brezhnev, who was then sort of in charge of the Kremlin, had another thought: the Yenisei and Lena rivers were even bigger, even farther away, with even more water. They would restore the Aral, giving it as much water as it had once received from both the Amu Darya and the Syr Darya combined. Despite dire warnings about the ecological consequences, this lunatic scheme, not unlike some of the more preposterous diversion schemes proposed by American water planners, was kicked around by Soviet bureaucrats for a decade or more, while the Aral continued its descent into obsolescence.

It wasn't until the Gorbachev era of glasnost that the scheme was put to rest. By then, Russian newspapers were vying with one another to find atrocious stories to tell, and the Green movement was springing up everywhere, easily finding fodder for its entirely justified indignation.

In 1988 the Central Committee decreed that cotton production was to be reduced so that the Aral could receive water in ever-increasing amounts, through 2005. Two years later, having seen that this remedy was not nearly enough, the Supreme Soviet declared the Aral a disaster zone and decreed an emergency relief effort. But by the time the Soviet Union itself collapsed a few years later, nothing much had been done. The water diversions had been marginally reduced, but that was all. By 1994 Moscow was officially out of the game, and the Aral crisis was now in the hands of the five brand-new Central Asian nations. The Amu Darya and the Syr Darya had become international waterways, which in turn created an urgent need for international cooperation to deal with the environmental degradation.

The economic numbers are not trivial. Water diverted from the Aral still supports agriculture worth billions of dollars, employing millions of people. Although Uzbekistan is as far north as Maine in the United States, and probably would never have grown cotton had it not been for those Soviet planners with their maps and rulers and impenetrable ignorance, cotton is Uzbekistan's chief crop, making up one-third of all exports, and the country's leading hard-currency earner. Agriculture accounts for 43 percent of gross national product and employs one-fifth of the population. Turkmenistan, for its part, depends on agriculture for 46 percent of GNP. In Kazakhstan the figure is 33 percent, with agriculture employing one out of every five Kazakhs. The World Bank estimated in 1998 that the total population of the five states would more than double to 86 million by 2025, before stabilizing at some later date at 186 million. There is not enough water to handle more than a fraction of the agriculture required to feed all those people.

In 1992 the five Aral Basin states signed an agreement pledging cooperation, using as a measure water quotas allocated under the Soviet regime. Little, however, was done. The following year, with World Bank support, they signed the General Agreement on the Aral Sea Crisis in the town of Kyzyl-Orda. Russia took part in this meeting as an observer, but never promised any money; no one believed it had any, and the routine demands that were made were halfhearted. Another meeting, in 1994, resulted in a variety of offers to reduce water consumption. Again, nothing much happened, except that Turkmenistan surreptitiously began to extend the Kara Kum Canal toward the Caspian Sea to the west, sneakily extending its own farmland.

In June 1994 the five states met in Paris to see if they could hammer out a solution. An unstated but important objective of this meeting was to "alert the international donor community [the rich countries] to the Aral disaster." Viktor Dukhovny, head of the Scientific Information Committee of the five-state Interstate Water Commission, remained optimistic that despite all the obstacles, progress would be made. In an interview in Paris in 1998, he laid out an "action program" that would begin around 2002 and terminate somewhere around 2040. This program sounded thorough and sensible — on paper. It involved an emphasis on permanent water conservation by all parties; the joint management of transboundary water resources "based on national parity of rights and responsibilities"; the systematic development of a regional water database; the involvement of the public and NGO community; the gradual shifting to a more user-pay system of water distribution; and the popularization of water conservation among the general public.[1] It all depended, he admitted a little glumly, on "a constant consensus among participants." Whether that was likely he declined to say.

The Aral continues to shrink and may soon be entirely lifeless. Even if the five states succeed in stabilizing water withdrawals, this restraint would do nothing to deal with the pollution problem. Full restoration would require wholesale changes. Agriculture would have to be deemphasized, for one thing — a possibility, since the region has oil and gas to spare and is urbanizing rapidly. With World Bank help, the remaining irrigation could switch to less water-demanding crops and itself become more efficient. Israeli agronomists experimenting on an Uzbek cotton farm claimed that they had increased yield by 40 percent while reducing water consumption by two-thirds. The Israelis had a self-interest in being there: to observe firsthand how to cope, or not cope, with an overstressed resource. They have an eye to their own future, as the Dead Sea shrinks and the Sea of Galilee comes increasingly under stress.

But no one believes this full restoration will happen. The primary obstacle is money. The Central Asian republics are the poorest of the former Soviet Union and have little cash to invest in rebuilding irrigation infrastructure. The GNP per capita ranges from $2,030 in Kazakhstan to only $980 in Uzbekistan. Instead of being scaled back, agriculture is in fact expanding. Everyone is squeezing as much water from the system as possible. Competition replaces cooperation, and angry noises are emanating from all five capitals.

Tajikistan, where both rivers rise, is the key jurisdiction. It has plenty of water, but its own water is filthy, and corruption in the capital, Dushanbe, is endemic. In 1997 a typhoid epidemic killed more than a hundred people and put nine thousand in hospital. Water treatment plants are in disarray, with no money even for chlorine, while the fleet of new limousines for government cronies continues to grow. The state government has so far refused to vote the 1 percent of GNP it is supposed to put into solving the Aral crisis; on the contrary, as the upstream owner of the water, it is demanding payment from downstream users for "its" water, particularly from the Uzbeks, traditional rivals. As a result, the Aral Fund is completely empty. The Tajiks point out that the Uzbeks don't scruple to charge them for the natural gas they supply, and will turn off the taps if payment isn't prompt. Why one rule for gas and another for water?

Uzbekistan has troubles of its own: a catastrophic thaw in a glacier in mid-July 1998 produced flooding that destroyed a dam on the Kuban-Kel Lake, causing it to collapse; forty-three people were killed. The Uzbeks, who are entirely dependent on Tajik water, are in turn accusing the Tajiks of exporting pollution and disease in the guise of fresh water. There have been threats from radical Tajik nationalists to deliberately poison the water leaving the country; and on three occasions Uzbek water bandits crossed the frontier to steal tankerloads of water and to dynamite Tajik storage facilities. Posses have been seen patrolling the borderland irrigation canals on both sides to combat water poaching. Outright fighting has already occurred between Kirghizis and Uzbeks, Kirghizis and Tajiks, and Turkmen and Uzbeks. The Uzbek president, addressing the Central Committee of the local Communist Party in 1998, admitted that "we have already had dozens of conflicts over land and water." A running battle in the Osh region of Kirghizstan between resident Uzbeks and Kirghizis killed almost three hundred people.

The Uzbeks have also admitted drawing up a contingency plan to invade northern Turkmenistan. Conflict specialist David R. Smith, writing in *Post-Soviet Geography,* identified two possible immediate war zones. The first zone is Kirghizstan and Uzbekistan, "given Kirgistan's riparian position as an upstream country and Uzbekistan's great interest in access to its water." The second zone is the two downstream countries, Uzbekistan and Turkmenistan, in which there have already been sporadic outbreaks of local violence against perceived water diversions.[2]

The 1992 conference that allocated quotas to the five states also set up an international body, the Interstate Coordinating Committee for Water Resources, which meets five times a year to reexamine water quotas. This commission has in turn established what are known as BVOs, the Russian acronym for River Basin Commission, which in theory have supranational powers, and therefore constitute the basis for international agreement. The BVOs have supposed control over water resources across state borders, including canal headwaters. A 1998 World Bank report was positive about the credentials of the people running the BVOs and the seriousness of their intentions.

But the BVOs, too, are being starved of the money they need to do their work, and meanwhile the sea, and the people who live around it, continue to suffer. In practice, planners have reduced the amount of money to be spent on water management sevenfold since Soviet times. The Karakalpaks, hardest hit by the crisis, declared their sovereignty from Uzbekistan in 1990 and began to set up their own water distribution regime. No one took either action seriously, since Karakalpakstan has even less money than everyone else in the region.

"We probably only have ten years left before the sea is beyond hope and the whole region is turned into desert," a Karakalpak bioecologist, Akmed Hametyllaevich, told a researcher from the Green organization People & the Planet. A local physician put it even more starkly, saying, "We may very well be witnessing the death of a nation as a result of human folly."[3]

7

To Give a Dam

*Dams are clean, safe, and store water
for use in bad years, so why have they
suddenly become anathema?*

EARLY IN 1997 the people who build dams, the people who
fund them, and the people who hate them got together in
Switzerland to see what could be done about them. The co-
sponsors were the World Bank, which had put up the money for
many of the huge dams in developing countries but which, under
the presidency of Jim Wolfensohn and the hidden guiding hand of
Maurice Strong, had been turning skeptical; and the International
Union for the Conservation of Nature (IUCN), usually known as
the World Conservation Union, a network of conservation agencies and organizations whose views on dams had always been a
model of transparency. To the IUCN, dams were almost always
wholly bad. In a perhaps subliminal slip, an Internet press release
issued by the IUCN after the meeting had misspelled "dam" as
"damn."

The subject matter of the two-day Swiss symposium was a review of fifty World Bank–funded dams carried out by the bank's
Operations Evaluations Department. Clearly the hapless officials
in Evaluations hadn't yet got the new signals being given off at the
top. Their review had stated that "the benefits of large dams far
outweighed their costs." This wouldn't do. As an announcement
issued later by one of the participants, the International Rivers
Network, put it in terse bureaucratese, "Participants at the work-

shop largely agreed that [the study] was based on inadequate data and flawed methodology."

There were a few more measured voices heard from, among them Aly Shady, a vice president of the International Water Resources Association and president of the International Commission on Irrigation and Drainage, whose paper asserted that "the survival of the human race in the next millennium will be tied to the success in managing fresh water. One cannot isolate large dams from the overall water management of the planet. Existing and new dams will continue to play a major role in the management system." In the end, everyone at the meeting agreed to set up an International Commission on Dams, made up of "eminent persons, acceptable to all sides," whose charge would be to develop guidelines for funding, constructing, and managing any dams that were still contemplated.

A series of self-congratulatory press releases followed. Patrick McCully, campaigns director of the International Rivers Network, said, "We are greatly encouraged that the World Bank and other dam builders have accepted that dam building has caused many problems, that there is a need for an independent review of dam impacts, and that existing practices through which dams are planned and built are in serious need of improvement." Shripad Dharmadhikary, a leading activist with India's Narmada Bachao Andolan (Save the Narmada Movement), said, "While we warmly welcome the willingness of the dam industry representatives to work with dam opponents in establishing an independent review, this does in no way mean that we will lessen the intensity of our campaigns against dams, for justice for dam-affected people, and for the implementation of equitable and sustainable alternatives." Peter Bosshard, secretary of the Switzerland advocacy group Bern Declaration, said, "We are delighted at the call for the establishment of an independent review; however, we are also aware that we will need to be constantly vigilant to ensure that the review is truly independent and that its terms of reference are as comprehensive as agreed."

The dam builders, for their part, maintained a glum silence.

Shortly afterward, McCully and his allies were back on the Internet, roundly condemning the World Bank for appointing all the wrong people to the commission and expressing, in the language of communiqués, their "grave disappointment." More silence followed, and then a third series of press releases, once again warmly welcoming the (new) commission.

What was going on? Why have dams — for so long accepted as a benign way to save water for a not-so-rainy day and to produce clean, nonpolluting power — suddenly become anathema?

Just off the roadway on the Nevada side of the highway that crosses the Hoover Dam into Arizona is a commemorative frieze and a monument to the men who died while building it. The monument is really to the heroic spirit of the age that produced the dam; the men who "fell," in the measly cant of politicians, are presented mostly to demonstrate the enterprise, courage, and true grit of Americans and to eulogize the skills of engineers. The structure of the dam is actually quite moving. It is made of brass and granite, concrete and steel; if you turn your head you can see the waters of Lake Mead lapping against the curving concrete of the dam and, on the other side, on the concave side of that sweeping mass of manmade stone, the dizzying drop to the Colorado riverbed more than 200 meters below. It's hard not to feel puny next to that immense artifact, and even harder not to feel awe. The dam is so massive . . . but also graceful, a sweeping curve across the canyon, even the 120-meter intake towers rising from stone niches carved out of the canyon and designed to look elegant and slender. Hidden from sight are tunnels blasted through the rock, first the diversion tunnels that created a new and temporary Colorado River while the dam was being built, and then the spillways, designed to handle twice the flood load of the river, each with huge brass doors incised with art deco devices.

Everything about the dam is improbable. It was built by a consortium of small-time contractors, most of whom were road pavers and railway track layers who had never built anything like this before. Not that anyone had built anything like this before: the temporary coffer dam alone would have been one of the largest dams in the world had they bothered to leave it standing; the dam itself was filled with 66 million tons of concrete; the cooling pipes laid in the dam would have made a refrigeration plant that stretched all the way to the Pacific coast, more than 200 kilometers away (the dam would have taken a hundred years to cool without those pipes, so great was the pressure); and so on and so on. The contractors were mostly small-minded men with maniacal obsessions, with not an ounce of aesthetic sensitivity among the lot of them, yet they managed to build a massive structure that was also beautiful, with a sense of design that matched the glories of great cathedrals. And it was built by men earning four dollars a day, in the blinding heat of the hottest place in North America, in the depths of the Great Depression, in just three years. It cost less than $50 million.

The Hoover Dam is important in another way, too. Humans had been building dams for thousands of years, mostly from mud and stone, and often from wood. In the century before Hoover, many dams had been built across many significant rivers, but none so wild and apparently irresistible as the Colorado. The completion of the Hoover unleashed a flood, if that's the right term, of other massive projects. In Russia, Stalin, not to be outdone, began construction of the system of canals and waterways that was to reach from the White Sea to the Caspian and the Sea of Azov, and began planning to reengineer Mother Volga, which ran through Russia's heartland. When he had finished, the river was not really a river anymore but a series of gigantic stepped reservoirs, some of them hundreds of kilometers long, which came to be called the Volga Cascade. The same thing happened on the Yellow and the Yangtze in China, the Zambezi and the Niger and the Nile in Africa, the Paraná in South America, and in a binge of dam building elsewhere. One of the things that Hoover set in motion was a change in the character of the world's waterways, permanently al-tering the ecosystems of entire drainage basins, in some cases changing the local microclimate, and in at least one case, the Nile, permanently changing a flow pattern that had sustained civilization for five thousand years. The numbers are startling. In 1900 there were no dams in the world higher than 15 meters. By 1950 there were 5,270, two of them in China. Thirty years later there were 36,562, of which no fewer than 18,820 were in China. The binge is slowing now, mostly because there are few rivers left worth damming, and because the ecologists are now, at last, beginning to count the costs.

I've been collecting dams — mostly, I think, because of my grandfather's obsession with them. The rationales for building these mammoth projects include the production of hydroelectricity (a "clean" energy source, after all), flood control, navigation, and water storage — in wet periods for use in dry ones, for irrigation, and for direct human consumption. Drought insurance was my grandfather's overwhelming reason. I like to visit dams whenever I can because of it.

Not so long ago I stood — very carefully — on the lip of the sheer 200-meter gorge on the Zambezi River, below Victoria Falls. I peered gingerly over the crumbling edge. Far below, an eagle was cruising, hunting. And below that I could make out a yellow raft as whitewater thrill seekers went hurtling down the river.

There was a furious debate about that gorge in the freewheeling newspapers of Zimbabwe and Zambia, which share it as a national frontier. A consortium of engineers had proposed the construction of a dam about 20 kilometers downstream. The dam would flood the gorge but not affect the falls themselves, one of Africa's greatest sights and a major tourist draw for both countries. It would, however, do away with the famous rapids downstream from the falls.

Green opposition to the dam centered on the nesting sites of the great eagles, which would be destroyed by the reservoir. This focus was a tactical mistake, I thought then, for it trivialized the debate, allowing opponents to dismiss protesters as people who would oppose something grand on the grounds that it would disrupt a few bird's nests. The region was hardly short of eagles, in any case. And, indeed, the press in Zimbabwe was having a field day, laying the sarcasm on thick. Zimbabwe wanted the dam. The other Zambezi River dam, the one 400 kilometers downstream at Kariba, had been built at great cost to both countries, but most of its hydropower had gone to Zambia. Zimbabwe wanted this one, although it had zero chance of being built. Neither Zimbabwe nor Zambia had the money, and the World Bank was less enthusiastic about funding large dams in developing countries than it had been, having become skeptical of their real value, and having better uses for the vast sums of money they inevitably consume.

I had seen the dam at Kariba a few years after it was built. Not quite Hooverian in scope, it was big and impressive enough, 128 meters high and more than 600 meters along its curving crest. Lake Kariba was nearly 300 kilometers long and some 30 wide, 100 meters at its deepest point, with lush shoreline and dozens of islands. These islands had become famous, or notorious, as the reservoir filled in the late 1950s and "Operation Noah" became a worldwide alert for the thousands of stranded animals that would otherwise have drowned. Independence movements at the time — for these were still colonies, white-ruled under the short-lived and widely hated Federation of Rhodesia and Nyasaland — pointed out snidely that more concern had been expressed in Europe and the Americas for elephants and leopards than for the 50,000 Tonga-speaking people who'd been forced to evacuate without an airlift.

By the 1990s, ecologists were divided. The natural habitat of millions of creatures had been fatally disrupted, but Lake Kariba had become home to a hugely successful population of hippos and other animals, and it provided reliable water for the wildlife as well as the farmers and residents of two countries. Tilapia, a species of fish alien to the Zambezi, had been introduced to Kariba and had flourished, driving out some native species and pushing others to extinction, yet making possible a brand-new fishing industry that employed thousands of small-scale fishermen and yielded more than 20,000 tons of fresh protein annually and sustainably. On balance, I thought then, Kariba was probably beneficial. It drowned some farmland, forced people to move, changed habitats, and permanently altered ecosystems. But it also provided clean power for industry and cities, and made possible the cultivation of land that had been barren.

Of course, that was before ecologists began to understand the true economics of dams.

There's no doubt where ecologists stand on the Aswan High Dam, on the Nile near the Egyptian border with Sudan. In their view it is an unmitigated disaster. A product of overweening ambition and cold war competition, it has damaged far more than it has benefited, and has placed the Egyptians in a policy bind: they cannot destroy the dam, but neither can they tolerate it (though it's fair to say that this view is not shared by the people who manage the dam or by many of the people who use its water).

If the Aswan High Dam is indeed an ecological disaster, its worst perils are hidden. On Lake Nasser, the 600-kilometer-long reservoir formed by the dam, everything looks peaceful. The feluccas ply the waters as they have done along the Nile for forty centuries, identical in style to the boats depicted on the frescoes at Luxor and Karnak. You can sit by the Nile at Aswan of an evening, and if you're lucky, you may see the feluccas drifting by in the light of a blood-red tropical moon. But below the surface, and in the lands north of the dam, problems are accumulating.

There had been an earlier dam at Aswan, completed in 1902. At the time it was one of the largest dams in the world, more than 2 kilometers long and pierced by 180 sluices. Those sluices, and the fact that it wasn't longer and higher, meant that the dam's net effect was relatively benign. It held back about 4 million acre-feet from the tail end of the annual Nile flooding for later distribution, which was a benefit in dry years, but it passed on the bulk of the flood, with its critical load of silt.

The High Dam was different, a lot bigger and a lot longer: 100 meters high and nearly 4,000 in length. Gamal Abdel Nasser had skillfully played off the Soviets and Americans to get the funding he needed for this grandiose scheme (more than $1 billion, twenty times the cost of the Hoover Dam), and then persuaded the United Nations to pay for the shifting of the Great Temple at Abu Simbel, Rameses II's monument to his own deification. The moving of this monument out of the way of the waters of Lake Nasser rivaled in its engineering skills the original conception. The dam, at least in the view of its builders, was one of the world's great engineering works. The problem, however, is that the famous Nile silt doesn't pass through the dam. It stays there, slowly accumulating.

It's true that the dam has virtually eliminated the potential for flooding, thereby preventing damage to buildings, farms, and roads. It's also true that it preserves water for those years in which the Nile barely stirs at all. But Egypt was able to sustain agriculture for so many centuries, unique among irrigating cultures, entirely because of the annual deposit of fresh silt from the Ethiopian highlands and the marshes of Sudan. Now that same silt is piling up behind the massive walls of the Aswan High Dam — slowly but surely making it obsolete. The nutrient-rich ooze no longer makes its way to the Mediterranean, and this has led to the collapse of the sardine fishery in the Nile Delta — a collapse that more than offsets the new fishery introduced with such fanfare into the reservoir behind the dam itself.

The changes to the system's water deliveries has meant that the shoreline of the delta is eroding at about 3 meters a year. Egyptian farmers, now able to irrigate all year round instead of waiting for the flood, have been protected from the periodic droughts and famines to which the Nile system was so prone. But another consequence of intensive irrigation has been a steady rise in the salt levels of the soil, a rise so critical that Egypt hired an international team of engineers, including the former commissioner of the U.S. Bureau of Reclamation, Floyd Dominy, to advise it what to do. The panel's only suggestion, after it had finished shaking its collective head, was a radical and improbably expensive drainage system under the irrigated lands.[1] No one thinks this is financially practical, though everyone accepts it as necessary.

As the Egyptologist Peter Theroux puts it: "For more than 30 years the Aswan High Dam . . . has kept the river from flooding and depositing renewing sediment at its mouth. The delta has instead been inundated with catastrophic superlatives: it is among the world's most intensely cultivated lands, with one of the highest uses of fertilizers and highest levels of soil salinity."[2] But the delta is slowly sinking, a condition that will affect the groundwater. Egypt's precious soil is being buried beneath the concrete of Cairo's relentless sprawl. The coast is eroding, and chemical pesticides are killing marine life.

At the same time, the irrigation channels and the new permanence of the water have led to an epidemic of bilharzia, with infection rates approaching 100 percent in some areas. Egypt is not alone in finding that dams cause disease. Bilharzia and malaria have increased after the building of every large reservoir in tropical regions. But that's just one of the problems associated with dams.

In earlier years, when dams were still regarded as ecologically benign, the commonsense view was that their most serious problem had been the displacement of people. Yet dams cause worse problems than that. Ironically, it has been reliably shown that malnutrition often follows dam building instead of the expected bounty resulting from newly irrigated lands. The reasons are complex but stem to some degree from the failure of engineers to understand the ecological benefits of flooding and the positive effects annual floods have on ecosystems. In many regions the local people developed over the centuries an intensive and varied flood-dependent agriculture. After dams are built and irrigation follows, the broad and reliable array of local foods is replaced by irrigated mono-culture, often aimed at export markets. Diversified agriculture has been found to produce more per hectare than irrigated agri-culture.[3]

Dams also alter the flow, and sometimes the temperature, of rivers. They almost always lead to elevated salt levels in the sur-rounding soil, sometimes infecting groundwater. The change in sediment loads affects major resources, as the Aswan High Dam did the Nile Delta's sardine fishery. The same thing has happened elsewhere. Dams on the Niger decreased catches by one-third. In the Pacific Northwest, even after decades of retrofitting and tinker-ing, the number of wild salmon returning to the Columbia River is less than 6 percent of what it had been before the dams were built. In those reservoirs where new species are introduced, catches al-most inevitably drop, sometimes after an initial surge.

Dams also radically affect the nutrients and the saline mix in downstream deltas. Michael Rozengurt, a Russian aquatic ecolo-gist, has estimated that estuary areas can withstand alterations in their nutrient fluxes to about one-quarter, after which catastrophic changes take place. In south Florida the fish catch has been re-duced to 42 percent of its former level. In San Francisco Bay about four-fifths of the migration and spawning areas of salmon, shad, and striped bass have disappeared. The Snake River's coho salmon run is extinct. Most of the fisheries in the Black Sea, the Sea of Azov, and even the Caspian are near death.

In northwestern Canada, the W. A. C. Bennett Dam, a monster on the Peace River, is fatally disturbing fisheries in the Peace-Slave-Athabasca Delta, a World Heritage site and one of the largest fresh-water deltas in the world. The Athabascan Chipewyan First Nation complained to the Indian Claims Commission that their way of life had been damaged, and in a trenchant judgment the commis-sion had this to say: "As is glaringly apparent from the evidence in this case, it is more than the First Nation's treaty rights to hunt, fish, and trap for food that have been affected; the First Nation's very way of life and its economic lifeblood were substantially damaged as the government of Canada, armed with full knowledge of the ecological destruction that would follow, did nothing."[4] Reacting to an earlier complaint, the Canadian government had weakly maintained that it had no power to affect what the provincial government of British Columbia got up to, a notion that was contemptuously dismissed by the commission.

The sheer volume of water in large dams — and therefore its weight — can also set off man-made earthquakes. This is one of the worries about China's mammoth Three Gorges Dam, which is located in an area of seismic faults. There were also fears that the Daniel Johnson Dam on Quebec's Manicougan River, which created a reservoir of 141 cubic kilometers of water, the seventh largest on earth, would set off quakes. A much smaller reservoir on the same river did set off a "seismic event" of 4.1 on the Richter scale.

And then there is silt. Dams — all dams — silt up, for dams have a natural lifespan, depending on the silt level of the rivers they contain. In certain rocky areas with poor soil, dams will last for a long time, perhaps a thousand years, though no one knows for sure. In other cases, such as the Nile at Aswan, rivers that once ran brown with silt are now running clear beneath the dams. The retained silt, which is really just mud, is accumulating at the dam wall — millions and millions of tons of mud. For a generation or two, and possibly longer, dredging and silt removal can work, but in the end they get too expensive and the dams fill up, often much faster than the amortization rate for which they were planned. A true cost accounting for dams almost always shows the bottom line in red.

In parts of the world, careless agriculture — deep plowing of mediocre soil, clear-cutting woodlands to make way for fields, heavy grazing by stock animals, cheap fertilizers allowing fields to be worked year after year and thereby increasing erosion — have dramatically increased the rate of siltation. An extreme case, cited by Marc Reisner in *Cadillac Desert*, was the Sanmexia Reservoir in China, which was completed in 1964 and taken out of commission four years later. Reisner also cites the Tehri Dam in India, the sixth highest in the world, which had its lifespan reduced to thirty years because of deforestation in the Himalayas, and the Dominican Republic's Tavera project, completed in 1973, which by 1984 had accumulated silt to a depth of 18 meters. "In thirty-five years," Reisner wrote, "Lake Mead [behind Hoover Dam] was filled with more acre-feet of silt than 98 percent of the reservoirs in the United States are filling with acre-feet of water." This mud will never be removed. There is too much of it, and there is nowhere to put it even if anyone could afford to remove it.[5]

There are two more unforeseen consequences of heavy silting. The first is the destruction of deltas. The Nile Delta is the most obvious example, but the Mississippi Delta is under threat too. The silt that built up the Mississippi-Atchafalaya Delta no longer reaches the river mouth because it is being accumulated behind a series of dam walls. Eventually, unreplenished, the delta lands will wash away. The second problem is that silt-free water flows faster. River basins are scoured more thoroughly, becoming deeper and more dangerous. The floods of the Missouri and Mississippi rivers in 1993 and the Red River in 1997 were entirely predictable. As Janet Abramovitz put it in her WorldWatch report *Imperiled Waters, Impoverished Future,* such floods "provided a dramatic and costly lesson on the effects of treating the natural flow of rivers as a pathological condition." Humans had tried to wall the rivers in and had colonized their natural flood plains for farms and towns. The floods were an attempt by the rivers to reclaim what had always been theirs. The dams and weirs were not flood control but flood threat transfer mechanisms.[6] "What earlier would be a seven-day flood is now a seven-month flood because the water doesn't know where to go," was the way Anil Agarwal put it in 1998. Agarwal, the director of New Delhi's Center for Science and Development, was reacting to a disastrous year in which more than seven thousand people were killed in Asian floods, on China's Yangtze and India's Rapti among others, events that were often thought to be wayward, capricious, and inexplicable, but which more and more were seen as a reaction to human-caused environmental changes, especially deforestation.

The Rhine is the most severe example of this process. It is no longer really a river but an engineered waterway of levees, concrete embankments, locks, flow-control devices, hydro plants, weirs, and channels. Once the Rhine meandered over an extensive flood plain; now, cut off from its natural controls, it flows twice as fast as before in a channel up to 8 meters deeper, dramatically increasing the power of its floods. The Netherlands is forced to absorb the full effects of the Rhine's more powerful flow, largely caused by the Germans and the French, neither of whom has offered to pay for the damage the faster-flowing Rhine has caused. The Dutch have been diplomatically polite, although there are signs that their impatience is growing.

Sometimes, though, the imperatives of politics are on conservation's side. In 1994 the National Wildlife Federation and the state of Nebraska won a $7.5 million lawsuit against dam builders on the Platte River. The money was used to purchase and restore river channels and their surrounding wetlands, allowing the trust to pick up about 4,000 hectares of land, including more than 20 kilometers of river frontage.

Dam collapses are not exactly unknown. Chinese government reports indicate that in 1975 alone, one of the wettest years on record, dam collapses killed a quarter of a million people and brought famine and disease to more than 11 million others. Even in the United States, a country that prides itself on its engineering skills, dam collapses occur. The Teton Dam on the river of the same name collapsed in 1976, even before it was completely built, wiping out three towns and hundreds of thousands of hectares of farmland, which the flood scrubbed down to bare rock; fortunately there was enough warning that there were only a handful of casualties.

At the Glen Canyon Dam on the Colorado River in Arizona, a

section of spillway was destroyed by floodwaters in 1982. The spill-way runs directly through the sandstone underlying the dam. Philip Fradkin, in *A River No More,* described the event this way: "Two thousand tons of water per second soared from the spillways and the river outlet works on both sides of the dam in the most spectacular display of cascading water ever seen on the artificially controlled river. Rumbling noises were heard on June 6. The jets of water became erratic. Water the color of diluted blood, chunks of rocks, and pieces of concrete issued from the mouths of the spill-way tunnels, as if the dam was mortally wounded." The level of the reservoir peaked, Fradkin points out, at 3,708.34 feet above sea level on July 15 — six-hundredths of a foot below the point where engineers forecast control might have been lost. Afterward, the Bureau of Reclamation declared to its satisfaction that the dam it-self was never in danger, though written documents at the time showed that bureau engineers "feared for the safety of the dam and its foundation." Should we be reassured by the embedding in the structure of "1,142 strain meters, 264 joint meters, 74 resis-tance thermometers, and 60 stress meters"? Or that to "measure the dam's movement," five plumb lines were installed inside the dam, "each consisting of a 637-foot stainless-steel wire holding a 26-pound weight immersed in a barrel of oil so as to damp its movement"?[7]

British Columbia's W. A. C. Bennett Dam, one of the world's largest earth-filled structures (it is 2 kilometers across the top and 183 meters tall; its Williston Reservoir covers 166,000 hectares and holds 70,308,930,000 tons of water), had some uneasy moments in 1997 when two large sinkholes were discovered just downstream. Emergency repairs were undertaken, and in 1998 B. C. Hydro, which operates the dam, spent $4 million reinforcing its drainage system by gouging out a 5-meter-deep trench and backfilling it with rock. Ron Fernandes, Hydro's area manager, assured everyone who would listen that the dam was perfectly safe. "Geophysical tests confirmed the success of the sinkhole repairs last year," he said, "and we plan on conducting these tests on an annual basis as part of our ongoing surveillance program." It was unclear whether this assurance — and the closing of the road across the crest of the dam in the summer of 1998 "to install additional monitoring instrumentation" — made the people living downstream feel better or worse. It's undoubtedly unfair, but when a bureaucrat talks about "ongoing surveillance," the assurances of politicians come to mind, and the tendency to skepticism can be irresistible.

If huge dams are such a menace, why are we still building them? Sometimes, as with the Okavango pipeline, need trumps caution. Sometimes there is little choice.

Up in the high hills of Lesotho, the Worlds Without Horizon as the Basotho call them, the bulldozers and earth movers are scraping away the brush, soil, silt, and shale to get down to bedrock, hard granite in which they can anchor the first of a series of dams that will, when it is finished, be the Highlands Water Project. The World Bank is funding this scheme, one of the last of the massive dam projects it has agreed to support, and every few months an anxious delegation flies into Maseru, the mostly sleepy (if occasionally riot-torn) little Lesotho capital, to see how things are going.

So far they seem to be going well, but there have been protests from the villagers whose homes will be flooded by the new reservoirs and whose farms will be submerged. Lesotho does not have a lot of farmland. The country is largely mountain, austere and beautiful, wonderful for hikers and possibly shepherds, though no paradise for farmers. But Lesotho's struggling government has had to make hard choices. The country has nothing to export but its labor, and half the country's men are away at any one time, most of them working the gold mines in Johannesburg. The new South Africa, however, is a more militantly nationalist South Africa, operating under the notion that jobs should go to South Africans first. If that happens, Lesotho's unemployment rate will head up past the 60 percent range, and Maseru's one casino won't be enough to offset the disruption. This was one reason why unrest spread across the tiny kingdom late in 1998, leading to a bungled South African military incursion.

One other thing Lesotho has in abundance is water. It rains a lot there. The hot air sweeping in across KwaZulu Natal from the humid Indian Ocean coast condenses as it crosses the Drakensberg Mountains, and rain falls in sheets in the hills. Why not build a series of dams to catch the water, install generators, and export both electricity and water to ever-thirsty South Africa? I thought this a good scheme when I first heard it. My grandfather's parched farm had been no more than 100 kilometers from the Lesotho border, and he would have paid most of what he had for a reliable supply of water. No doubt the dams would fill quickly. No doubt little would be lost. Lesotho would never be able to feed itself anyway, and since it was going to have to import food, it might as well have the hard cash to pay for it. Engineers had said the water in the hills was almost silt free; the dams would last for hundreds of years.

Traditionalists opposed the scheme; not only would villages be flooded but burial grounds too. There were a few violent demonstrations and at least one gruesome ritual murder that the news media quickly suppressed. The World Bank grew ever more anxious about its investment. Traditionalists making common cause with environmentalists had opposed other dams, successfully, in Nepal, on the Narmada in India, and elsewhere. In 1997 the Malaysian government finally announced it was canceling the $5.5 billion Bakun Dam, which had been strongly opposed by environmentalists who insisted the dam would be a financial as well as an ecological disaster. Patrick McCully of the International Rivers Network had been exultant: "The Bakun cancellation highlights the economic unviability of large hydro projects. Bakun was the dam industry's flagship for privatized hydro projects — and the flagship has sunk." The dam's contractors, a German-Swiss-Swedish-Brazilian consortium, maintained another in the series of glum silences to which engineers were becoming accustomed. Often, time had proved the traditionalist-environmentalist alliance right, even when they lost. And did a worsening economy in South

Africa, where the rand collapsed in 1998, mean that no one would be able to buy Lesotho's newly produced electricity? The World Bank was right to be cautious.

Nevertheless, at the end of 1998, the project was proceeding. It was inevitable. There was, in essence, no choice.

Across the world, literally and in sensibility, another huge project was taking shape, this one with neither the advice nor the money of the World Bank. The Chinese were financing the Three Gorges Project entirely on their own.

This project has set off an uproar among international conservationists, ecologists, political dissidents within China, and artists. The Three Gorges Dam, on the middle Yangtze, will submerge one of the most picturesque places on the planet, an inspiration to Chinese painters and poets for three millennia, and the site of many shrines and temples. It will end forever the extraordinary and extraordinarily difficult boat trip through the awe-inspiring gorges. The 600-kilometer reservoir will, not incidentally, mean the removal of 1.5 million people and the submerging not only of villages but of a major industrial city as well. The submerged spillway bays, twenty-seven of them, each with the average capacity of the Missouri River, are well beyond proven world experience. And this on a river prone to flooding and in a country where dam collapses, as we have seen, are hardly unknown, and in a region crisscrossed by several seismic fault lines.

Flood control is one of the primary reasons the Chinese give for wanting to build the Three Gorges Project in the first place. According to Chinese statistics, the Yangtze has seen 214 "major floodings" in the past two thousand years. In the last hundred years alone there have been four "disaster level" floods; the recurrence frequency is now about one flood every ten years, as deforestation and the draining of wetlands have made the river ever more powerful. The plains along its middle and lower reaches are heavily developed, with industrial cities and intensive farming. The river's own discharge capacity is only 70,000 cubic meters per second, but it routinely reaches 100,000 in high-water floods, and the dikes built between Jingjiang and Hunan are not always strong enough to protect against flood crests that can reach 17 meters above normal.

In 1998 floodwaters crested at 15 meters above normal and threatened to overwhelm the city of Wuhan and its 7 million inhabitants, and by the time the crisis was over, more than three thousand people were dead and millions more homeless. As many as 240 million people, fully one-fifth of China's population, had in some way been affected by the flooding, and Chinese economists figured that the flood would shave almost a percentage point off the GDP for the year. Even the normally docile provincial legislatures were vigorous in their criticisms of the policies — or lack of policies — that had brought the disaster about. Lu Youmei, the chairman of the Three Gorges Development Corporation, said rather tactlessly at the height of the crisis that "if the Three Gorges Dam had been built by now, the problems of flood control would already have been solved." The project's opponents, in contrast, maintained that the dam was partly at fault. So much money and so many resources were being diverted to it, they said, that the building of levees and dikes and the clearing of the riverbed had been neglected.

The other reasons for going ahead with the project, or the "benefits," as Chinese hydrologists prudently call them, are power generation, navigation improvement, aquaculture, tourism, ecological protection, environmental purification, development-oriented resettlement, water supply and irrigation, and the transfer of water from south to north — all of which, in the enthusiastic encomium of C. Yangbo of the University of Hydraulic and Electrical Engineering in Yichang, make it a "unique super project worldwide." This is the same shopworn list once trotted out by the builders of the Aswan High Dam and the propagandists of the U.S. Bureau of Reclamation. It is perhaps doubly ironic in this case that tourism should be included, since one of China's major tourist attractions will simply cease to be, and another ordinary-looking lake will take its place. As for "ecological protection," hardly anyone has figured out what that means in this case.

It is true that, downstream, there are 1.5 million hectares of fertile farmland, several major railway lines, and 15 million productive citizens, all of whom could use a little protection from flooding. And, of course, all are vulnerable should the dam ever fail. China's construction standards (see dam collapses, above) are not exactly reassuring. Nor is the fact that, as of mid-1999, ninety-five cases of "corruption" involving the Three Gorges Project had already been prosecuted, and a shoddy bridge had collapsed just upstream, killing forty people.

For the dam to be able to deliver on its promise of hydroelectric generation, the engineers would have to find some way of flushing out the silt accumulating behind it. So far, no plans have been formulated that would do this. Even if they succeed, the best calculations are that the dam's electricity would cost three times as much as electricity from other available sources.

The final reason for building the dam is the possibility of transferring water from the middle Yangtze to Beijing and the parched northern plains of China. But it will take nearly twenty years to get the water there, far too late to avert the looming crisis. And even then, will it have been worth its $29 billion cost?

Late in 1999, in fact, there were signs that powerful forces within the Chinese government had answered the question with an emphatic negative — Premier Zhu Rongji in particular. The gruesome despoiling of the sublime landscape might yet be averted. But, then, what of the conduits to the north considered so necessary by water planners in China? The north, as we shall see, is running out of water. Without water from the Yangtze, those northern shortages could become critical.

And then what?

8

The Problem with Irrigation

*Irrigated lands are shrinking, and
irrigation is joining dams on an
ecologist's hit list. Why?*

PESSIMISTIC PROJECTIONS in the middle of the twentieth century often indicated that the world would run out of food
within decades — that agriculture could not possibly keep up
with explosive population growth. Instead of global famines, however, we had the Green Revolution, a biped with one foot in chemistry (pesticides and fertilizers) and the other in managed water
distribution, or irrigation, which together caused a quantum leap
in agricultural efficiency and productivity. As a result, India and
China, the focus of much of this Malthusian angst, not only fed
themselves but contributed to world surpluses as well, selling grain
on the open markets. India, particularly, seemed bent on irrigating
every possible hectare — 113 million at last count.

But now the per capita extent of irrigated land is shrinking on a
global basis. Why is this?

The first irrigation scheme I ever saw was very small, a single
narrow furrow on my grandfather's farm which led from the corrugated iron storage tank out back of the house to a patch of
ground in front. It meandered a bit — the workers had skirted
rocks rather than move them or break them up — and it wasn't
more than a third of a meter wide and about twice as deep. Narrow, yes, but wide enough for my sister, little more than a baby, to

plunge in one day. I still remember the shrieks and the running farmhands, and how one of them plucked her out just in time, seconds before the swimming snake got to her, a poisonous adder, its sinuous body driving it through the water, straight as an arrow from a Bushman hunter. She got out in time, but for a long while afterward I dreamed of that evil black head striking through the water, its tongue flicking, a creature with a sinister purposefulness. Or so I thought then. Later, when I was older, another thought interested me: Where did the damn thing come from? In this dry and arid land, how did a water adder find its way to water that had never been there before?

The corrugated tank was fed by a clanking windmill. Alongside a third of it was a concrete trough for the cattle to drink from. A stopcock along the side filled the furrow, and the furrow led to a small strawberry patch in deep and well-drained soil close to the house. My grandmother never persuaded her husband to take the water to the house itself: all her life she was obliged to fetch her own water for cooking, cleaning, and washing. My grandfather — my Oupa, as we called him in Afrikaans — believed that water was too precious to waste on taps in the kitchen. Water was for the crops; humans could fend for themselves.

The strawberry plants were watered directly, by bucket. There were no sluices in the furrow: there was not enough water for that. Every day the farm workers gave each plant exactly what it needed, and twice a year my grandfather could hitch up his wagon and take the crop to the farmers' market in town, five hours away by horse-drawn cart, whence it would go by train to the provincial capital, Bloemfontein, 100 kilometers away. What happened to it there I never knew. We also ate the berries at home, fresh and sun-ripened and watered by moisture from the center of the earth, and we drowned them in fresh Jersey cream from Oupa's herd, perhaps after a meal of home-grown roast mutton and asparagus spears from the bed next to the strawberries. We never thought much about it. It was just how we lived.

Oupa never learned to read and write. He was a Boer soldier

fighting the British when he should have been in school, and he remained unlettered all his life. But he was not a fool. He was a careful farmer, with a scrupulous love for natural things, and through
an intuitive understanding of how plants grew he had devised the
very model of a modern irrigation scheme: his water was stored in
a tank, not an open dam; the furrow was lined with stone and
filled only when needed; the plants were watered directly; and the
soil was properly drained.

The manuals at the agricultural colleges now recommend
schemes very like it. More massive, of course — sometimes hugely
more massive — but in essence the same, closed systems with little
wastage. It was, in all the best senses of the word, a sustainable
practice. And just as important, the good drainage and careful watering meant there was little chance he was poisoning the soil with
increasing salinity.

Much bigger farmers, and much more sophisticated farming
operations, were not paying the same attention.

Irrigation is an ancient technique, going back five thousand years
in Central Asia and a great deal further in Mesopotamia. It was
used in imperial China, in old Laos, in ancient Africa, in Tanzania
and Zimbabwe, and in Old America, before the Spanish ever laid
eyes on the place. For millennia, farmers have used aqueducts and
ditches to take water to where it was needed, and furrows with
crude sluices to divert the water to their crops, giving them increased yields and lessening their dependence on the caprices of
the weather. But on any real scale, irrigation is not much more
than a century old; it needed the massive dams and water diversions of the modern era to make it possible.

The numbers are revealing: two hundred years ago total irrigated land probably amounted to as little as 6 or 7 million hectares, not much bigger than Long Island, New York, or metropolitan Los Angeles and its exurbs. By 1900, the total had jumped to
some 50 million hectares, and that number nearly doubled in the

half-century that followed. Of the estimated 230 million hectares currently under irrigation, half were added since 1950, and dozens of countries, ranging from little Israel to massive China, rely on irrigation for much of their domestic food production — Pakistan up to 80 percent. Most of North America relies on rain-fed farming, but without irrigation, agriculture in California's Central Valley, one of the world's most productive vegetable and fruit resources, would hardly be possible. Only 15 percent of the world's cultivated land is irrigated, but irrigated land accounts for almost 40 percent of the global harvest. Without irrigation, yields in the world's major breadbaskets — on which the feeding of the planet is dependent — would drop by almost half.

But this rate of expansion couldn't go on: there was only so much water available, and by the 1980s the rate of construction of new dams was slowing. All the easy rivers had already been dammed, for one thing, and the expense had escalated dramatically, to the point where many of the proposed new irrigation schemes had priced themselves out of possibility: they would have delivered water at a price not much cheaper than what the farmers could get for their commodities. In parts of Africa the cost of delivering water from some of the irrigation schemes on the drawing boards in the 1980s was projected to reach more than $10,000 per hectare, and, as Sandra Postel has pointed out, "Not even the double-cropping of higher-valued crops can make irrigation schemes at the top end of this spectrum economical."[1] In fact, she says, for at least the past two decades the population has been outstripping what irrigation increases there are. Projected population growth rates for the next thirty years will require an increase in food production equal to 20 percent in developed countries and 60 percent in developing countries to maintain present levels of food consumption. Yet per capita irrigation peaked at 48 hectares per thousand people in 1978, then dropped 6 percent in the following decade to 45 hectares per thousand people. The trend is accelerating.

The situation is unlikely to reverse itself, even if commodity prices rise high enough to catch the attention of water engineers. The costs of damming the remaining undammed rivers is prohibitive.

But the problem is worse than just a lack of expansion opportunities. Not only is no more land being brought under irrigation, but also a good deal of irrigated land is being taken out of production. Salination of the soil, to a degree that inhibits farming, is spreading at the rate of more than 1 million hectares a year. In Pakistan alone, 2 million hectares have been decommissioned, the soil poisoned by high salinity, and farm yields are down 30 percent from historic high levels. Egypt is showing similar declines. In the Imperial Valley in California, more land is being decommissioned than commissioned, again because it is overly salty. Improved drainage might save the fields, but there may not be enough water or money to do the job. Ironically, the Imperial Valley sits on an enormous geothermal aquifer which contains great quantities of salty water that could possibly provide a solution to the problem. Water temperatures in the huge underground pool rise as high as 280°C. Researchers are developing methods to use the steaming brine to generate electric power. They believe it is feasible that the clean wastewater produced by a power plant could be used to dilute the salty water in the fields.[2] In the United States as a whole, yield reductions owing to salinity occur on an estimated 30 percent of all irrigated land. Worldwide, crop production is limited by the effects of salinity on about half of the irrigated land area, and nearly two-thirds of the total area needs expensive renovation just to keep working.[3] The consequences for world food production are still unknown but are sure to be high.

That pesticides and chemical fertilizers have their own set of evil consequences is well known. That irrigation — the delivery of clean water to the plants that need it — brings another unexpected set of difficulties has come as a nasty surprise. But it is becoming more and more apparent that salinity and the increasing concentrations of noxious substances in water are the vulnerable underbelly of the irrigation revolution, threatening huge areas of other-wise productive lands. The problem is simple to state, though devastatingly difficult to fix: without great care and skillful man-agement, irrigation almost inevitably causes waterlogging, deple-tion and pollution of the water supply, and rising salinity in the soil. Left unchecked, these problems can eventually kill the soil altogether.

The first problem is one of concentration. Because water is such an efficient solvent, it picks up salts and minerals from the soil on its journey down from the hills. When it collects in the shallow lakes in the valleys, salt pans to be, or when it is applied to irrigated land, some of it seeps into the ground, and some of it evaporates.

If the soil is well drained, the water that seeps downward might eventually trickle into an underlying aquifer, in which case it causes little harm. But if the soil is not well drained, or is com-pacted clay, or there is no natural outlet for the water, it will be-come waterlogged and the salts will begin to accumulate. Over time, heavy concentrations of salts get closer and closer to the sur-face, eventually reaching the root zones of vegetation. For a while there is a struggle. Some plants are less efficient at handling salts than others, as early irrigating cultures quickly learned. But even-tually even the salt-resistant plants succumb and the last vegeta-tion dies.

Not all the water seeps downward. Much of the water evapo-rates. In incipient salt pan formations, for every 100 cubic meters of water, some 75 will typically be lost to evaporation. If the wind conditions are right, some of the salts will be taken up into the atmosphere too, through aerosols. But nine-tenths of the salts will remain behind on the surface, and the pan becomes ever more saline. Eventually, the salts in the root zones meet the salts on the surface, forming a crust, and the lakes will become perma-nently dead.

In many cases in the past, these dead zones have also become arid. Less water runs into them than before, partly through changes in the local climate, and partly by increased radiation of heat from land stripped of its vegetation.

The second problem is accumulation — concentration in another form. This, too, occurs in nature, but seldom to a troublesome degree. Rivers are naturally efficient flushing mechanisms, and the salts that accumulate in them tend to make their way to the sea. It's true that where groundwater runs through naturally occurring saline formations, the water that enters rivers might contain a higher concentration of salts, and it is also generally true that rivers running through naturally alkaline or saline soils tend to become saltier the farther they travel. Nevertheless, the concentration is usually not critical, and over the eons the local wildlife will have adapted to it, to such a degree that many delta fish habitats have become dependent on specific salt concentrations.

But where river water is taken up by humans for use in irrigation, the method of concentration — and the concentrations themselves — change. For irrigation, river water is either diverted to canals or directly taken up by pumping. Much of the water so diverted is lost in the process. As a working average, almost 60 percent of the water intended for irrigation never gets to the croplands. Leaking pipes, unlined canals, evaporation from open reservoirs and canals, and poorly directed spraying cause much of the water to be wasted. Some of this wastage returns to the groundwater, so it is not "lost," except in the sense of adding to farmers' costs. Much of this missing water seeps through the soil, picking up some of the already accumulating salts as it goes, and returns to the river farther downstream. There it is recaptured in another irrigation scheme, and the same thing happens. If the river is intensively managed, it will pass through several reservoirs before eventually making its way to the sea. At each stage, in each reservoir, water is lost to evaporation, as much as 2 meters a year, especially in dry climates, further concentrating the salts. In certain rivers water may be "used" more than a dozen times, in each cycle becoming more and more salty. By way of comparison, in parts of Spain, for example, where the soil is not particularly alkaline, river salinity can increase from around 400 to 600 parts per million, or less at the river's source, to somewhere around 1,200 to 2,000 parts per million before it discharges into the sea. In the American West, where the soils are poor to begin with and there is a good deal of natural salt, the intensifying power of the process is much greater, and salt concentrations can be almost twenty times higher at the mouth than at the source.

Even the most pristine water contains some salts, typically about 200 parts per million, compared with the 500 considered safe for drinking. (The ocean, by comparison, contains about 35,000 parts per million.) As Sandra Postel points out, applying 10,000 cubic meters of irrigation water per hectare, a fairly typical rate, would add around 5 tons of salt to the soil annually.[4] The salt is therefore being "stored" in the land and in river basins instead of being flushed out to sea. As a consequence, productive land is becoming salt-encrusted and barren.

None of this has to happen, and none of it means that irrigation inevitably poisons soils. Where drainage is good and the soils not naturally alkaline, carefully managed irrigation can persist for centuries without harm. In areas where irrigation is used as a water supplement rather than the whole diet, little harm is likely to befall the land.

It is where irrigation is intensive, the soils naturally poor, and the drainage either inadequate or nonexistent that the most serious problems will occur. The American West — where the Great American Desert has been turned into one of the planet's breadbaskets — is the most notorious example. In some areas in the San Joaquin Valley, the salts are so obvious that the earth looks as though it has dandruff; not even weeds grow there, and — shades of the Aral Sea — the salts drift on the winds, sometimes for hundreds of kilometers, exporting the problem to distant communities. Unless something drastic is done, and done soon, almost 1 million hectares of the most productive land in the world will be irretrievably doomed.

But what can be done? More efficient irrigation would help — the large-scale application of the methods my grandfather had worked out intuitively. The Israelis, always searching for better techniques, have improved efficiencies sixfold by using laser technology to get fields absolutely level, and by reusing surplus water from one crop on other, more salt-resistant crops. Israeli water engineers have taken "water stress management" of crops to a high art, and water only when absolutely necessary.

The other solution is expensive: to install master drains under agricultural valleys that have poor natural drainage of their own. These drains would take the unused water to a reverse-osmosis desalination plant somewhere downstream, where it would be scrubbed before being dumped back into the distribution system. By most estimates, though, drainage and desalination systems would cost several thousands of dollars more per hectare than the land is actually worth. So would most of the other solutions suggested by agricultural engineers, such as chemical filtration or detoxification by genetically engineered soil microbes.

The most notorious example of a conduit that was built with nowhere to go was the San Luis Drain, built to take wastewater from the Westlands Water Project in the San Joaquin Valley to . . . well, to somewhere. Unfortunately, the somewhere turned out to be nowhere, and the water was instead dumped into a man-made lake called the Kesterton Reservoir. For a while this seemed like a good solution. It even attracted millions of wildfowl, which began to use it as a stopover on their migration routes. To help the water authorities write off some of the cost of the drain, it came to be called the Kesterton National Wildlife Refuge. Alas, a decade after it first became a wild bird reserve, the birds started dying or developing grotesque deformities. A furious fight erupted between conservationists and farmers over whether pesticides had anything to do with the killoff. Biologists finally pinned the blame on selenium, a mineral not toxic in small quantities, which was naturally present in valley soils. Irrigation and poor drainage had, unknown and unseen, been concentrating the selenium into lethal dosages.

When my grandfather died, the farm was sold to some cattle barons from the Transvaal. They consolidated it into an agribusiness, running great herds of beef cattle in land that was marginal for the purpose. After twenty years they went broke and the land was abandoned. I went back to see it a decade ago, but I will not return. The old farmhouse where I had sat on the stoop with my Oupa, staring down the lane of bluegums into the imagined future, was now just a ruin. The stock sheds had crumbled. The corrugated iron tank was still there, but the windmill was broken. And the furrow that had taken the water to the strawberry beds had filled in. You could still see where it had been, a slight depression in the dusty soil, but the furrow itself was gone and, with it, the precious water. I squinted down where the berry patch had been and, for just a second, I thought I could smell the newly wet earth and see the ripe strawberries glistening from the moisture. But my imagination wasn't really up to the task, and after I watched a dust devil dancing beyond the broken barn, I turned and left.

9

Shrinking Aquifers

If water mines ever run out,
what then?

A SMALL CANVAS BAG of water, sweating in the heat, hangs from a hook in the cab of a huge truck, swaying as the vehicle lurches across the desert, its great tracks as wide as a house scraping across the gritty stone and drifting sand. From the cabin high above the sand, the driver and his companion see nothing but more sand and scrub, and great gouges in the desert floor where the earth movers have left their traces. If the driver were to look up from the treacherous road, he would still see nothing — nothing but the same sand and stone, stretching to the horizon. And if he drove to that horizon, there would still be nothing — nothing but the Great Emptiness for a thousand kilometers of the greatest desert on earth, stretching all the way down to Chad in the south and, far to the southwest, Agadez in Niger and the fading Saharan termini of Gao and Timbuktu.

The small canvas bag is cool to the touch, damp. The driver, Ahmed, wipes his hand across his forehead, leaving beads of moisture. He finds the bag reassuring. It holds 5 liters of fresh water. In fact, the water was trucked in from Tripoli, hundreds of kilometers to the north, but it could just as easily have come from the desert oases of Tazirbu, Sarir, or Al Kufrah. Just off the road near Tazirbu is a small well that has been known to travelers for centuries, the Birbu Atla, not much more than a meter deep, the clear water mesmerizing to a thirsty passer-by. It bubbles as you look into it, little bubbles of air coming up from . . . where? Travelers' tales often speculated, though no one knew for sure. In the past, camel caravans took fifty days to plod their way from Timbuktu to Marrakech on the western route, or from Cairo to Gao on the eastern, threading through a scanty network of oases, following the veins of water.

Ahmed was only 60 kilometers from the Birbu Atla now. He maneuvered his huge vehicle — a crane, really, a mobile crane capable of lifting up to 200 metric tons — until it was parallel to the gouge in the ocher earth that ran arrow-straight to the north, vanishing up the road to the coast, so many hundreds of kilometers away. On his back — Ahmed thought of the cradle of his immense carrier as his own flesh and blood — was a section of pipe. He lowered it gently into the waiting trench, and another machine, made just for this purpose, nudged it into place with a solid thud.

The section of pipe Ahmed was locating was 7 meters long and 4 in diameter. He knew something about this pipe. All the workers did, for it loomed large in their lives, as it had done for the four years since they first began laying it. There was no other pipe in the world like it. It was bigger, and better engineered, using the most modern of manufacturing techniques imported from the West and adapted on-site by the swaggering South Korean contractors. A whole factory had been built just to make this pipe, with its alternating layers of steel and concrete. When the sections were laid together, there would be no leakage and hardly any sweating; it wouldn't even feel damp to the touch. That was as well, for what it would transport was more important than anything else, more important than oil, although oil was paying for it in the end. This pipeline would be one tributary of what Colonel Qaddafi was calling, with no hyperbole at all, the Great Man-Made River. What it would transport was, in this dry country, the most precious thing of all: water.

The pipeline was to take water from the deep desert and carry it to the populated coast. The Sahara was giving up its last secret, something it had kept hidden for ten thousand years. Down there, deep down, was water — lots of it.

The Sahara is the greatest desert on earth, the apotheosis of dryness, but even on or close to the surface it is not entirely without water. Several major rivers originate outside its boundaries, flowing through or disappearing into the desert itself, impacting in various ways on its surface and groundwater. Some of these rivers rise in the tropical highlands well to the south. The Nile, of course, is one of them, meandering out of Uganda and Ethiopia and flowing along the desert's eastern border to the Mediterranean. Several rivers flow into Lake Chad, including the Chari. The lake is shrinking now, but gives up its water in two ways: through evaporation and by seeping northeastward to recharge underlying aquifers. The Niger River rises in the sodden hills of Guinea but traverses the southern edge of the Endless Desert before turning south to the sea. To the north, the Saoura and the Draâ flow from the Atlas Mountains, as do many other streams and occasional streams, or wadis. Within the Great Desert are shallow, seasonally inundated basins called chotts, or dayas, and many other wadis, most of them remnants of systems that existed in the distant past, when the Sahara was still verdant, but some still occasionally flow in the rare desert rainstorms. Tamanrasset in Algeria was destroyed by a flood in 1922, to the terror of its inhabitants and the bewilderment of the geographers. There is a network of wadis and remnant pools in the Tibesti and Ahaggar mountains; even the great sand dunes of the desert store water, sometimes in considerable amounts, and seeps have been known in desert escarpments. Saharan aquifers have been recharged in historical times; in the time of Europe's "little ice age" (from the sixteenth to the eighteenth centuries) there was heavy rainfall along the southern and northern fringes of the Sahara, and very likely in the desert itself.[1]

There are also bizarre holdovers from ancient times. An explorer in the deep Algerian Sahara near Tamanrasset came across a spring that led into a sluggish creek no more than a few hundred meters long, in which wriggling tadpoles and small fish could be seen. And the French academician Jacques Couëlle once dug a well in the desert near El Goléa, south of the Western Great Erg, and at 44 meters the men struck water, brackish and churning, but still water. The next day the crew returned with baskets and found themselves pulling out fish, "completely blind, with only a membrane where there used to be eyes; they have reproduced . . . surviving more than 10,000 years. . . . [T]here were hundreds of thousands of them. . . . [W]e cooked some of them and they were delicious."[2]

Water in serious quantities was first discovered in the 1970s, in the Al Kufrah region of Libya, far away in the southeast quadrant, where it converges with Egypt, Sudan, and Chad. It was an accident; the prospectors were looking for oil, not water. The massive oil fields in the Sirte Desert to the northwest were good for generations to come, it was believed, though it never did any harm to find some more. They found a gusher all right, but it was clear, clean water instead of black gold. A few more wells were test-drilled in the region and they all told the same story: underlying the desert, in the sandstone caverns and underground canyons, was an immense pool, which came to be called the Nubian Aquifer.

A few years later more water was found, this time to the west of the country in the Marzüg Basin. It was, the geologists reported, another massive aquifer, probably unrelated to the first.

These discoveries came at an opportune time. Libya was parched. It was, after all, 93 percent desert, and its renewable water resources were less than 200 cubic meters per person per year, about one-fifth of what the United Nations considers "water stressed," and below what all hydrologists think of as a water-critical state. The population was concentrated in two coastal regions, the Gefara Plains and the northeastern Benghazi Plains. In the middle, sparsely populated, are the oil fields; water shortages there have always imposed boundaries around possible development.

The growing population along the coast and the paucity of rivers meant that groundwater aquifers were used for drinking water and sanitation. Even the Libyans, notoriously reluctant to admit anything other than Triumphs of Socialist Construction, have confessed to severe overdrafting of their coastal aquifers. A 1997 report by Saad al-Ghariani, of the Department of Water Sciences at Al Fateh University in Tripoli, found that these aquifers had been exposed to "unacceptable levels of piezometric declines and seawater intrusions with disastrous environmental and socioeconomic impacts" — pretty straightforward talk for bureaucratese, a frank admission that many of the coastal wells had turned saline and become unusable. New water, then, was critically necessary.

The Libyans knew that there were only three options: radically expand the desalination of seawater along the coast; set up economic and agricultural zones in the newly discovered groundwater basins in the deep desert; or take the desert water to where it was most needed. The last was not, they knew, unfeasible: Hadn't the Americans already built massive interbasin water transfers, what with their California State Water Project and the Colorado River diversions? It went against the grain to admit that the Americans had done anything worthwhile, but these were, after all, useful precedents. And hadn't Turkey just proposed its Peace Pipeline, which would take Euphrates water all the way to the Gulf States, refilling the Jordan River and watering the deserts of Iraq, Syria, Jordan, and Saudi Arabia as it went? Of course, the American transfers and the proposed Turkish pipeline were different in one significant way: they involved reengineering rivers, and not the one-time mining of an unrenewable aquifer. But needs must as needs do.

To outsiders, the Great Man-Made River (GMMR) seemed mad, a product of senseless ambition combined with endless flows

of money. To build the largest civil engineering project on the
planet, to spend $32 billion to take water from the desert to the
coast, to create new agricultural zones, and to allow the population
to increase and industry to establish itself as a result of ready access to water — to do all this, knowing that in thirty years, or forty,
or perhaps fifty, the water would inevitably run out . . . It seemed
insane.

But what were the options, where the alternatives? A British
firm hired to do the analysis found, to its own evident surprise,
that the Great Man-Made River was the most cost-efficient of the
three possibilities. The average unit cost of transferred water, their
report said, would be about 25 cents per cubic meter. Not cheap,
but best estimates for seawater desalination still came out somewhere between $3 and $5 per cubic meter, even using considerable
amounts of cheap Libyan fossil fuels to do it. The third option,
moving the people and the industries, made no sense. There are
no seaports in the desert. And when the water ran out, they'd be
stranded.

Of course, that applied to the Great Man-Made River too. When
the water ran out, as it must . . . What then?

How much water is there? Geologists estimate that in the Kufrah
Basin alone there is something like half a million cubic kilometers
of usable water. Based on a transfer of some 40 billion cubic meters a year, the local water table would drop about 1 meter a year,
which would mean it would last for about fifty years. Then, of
course, it would be gone.

Nevertheless, by a series of interesting logical leaps, the Libyans
consider this a sustainable project. How? As al-Ghariani put it:

> The question of sustainability . . . depends on water production costs
> and management skills rather than available water supplies, which
> are apparently sustainable for hundreds of years even in the absence
> of natural recharge of the aquifers. Sustainability can be assured if

the transferred water is utilized in such a way as to provide the national economy with the means and strength that enable it to develop alternative water supplies when the GMMR sources become uneconomical to pump, or are exhausted altogether.

To bring this project off, Qaddafi turned to one of South Korea's notorious *chaebol,* or family-run conglomerates, before the South Korean economy collapsed and the acquisition-hungry *chaebol* began to unravel. The chosen contractor was the Dong Ah Construction Industrial Company, a sprawling empire of construction, tourism, transportation, and finance companies that built roads, nuclear power plants, hotels, and other projects in Korea and abroad. It was headed by the flamboyant Choi Won Suk, widely known in Korea as the "Thinking Bulldozer" or "Big Man" for his ruthless ability to manage mammoth projects. Dong Ah's chairman undertook the Libyan project with his usual directness, undeterred by a corruption scandal involving payments to South Korea's notorious president Roh Tae Woo. Choi was convicted of bribery, but his sentence was suspended. He had just been doing what all Korean businessmen did at the time: assuring himself a place at the trough.

After a decade of construction, Choi finished Stage One of the project at a cost of less than $4 billion — under budget and ahead of schedule. It was a 1,900-kilometer waterway, 4 meters in diameter, connecting a network of pipes and ditches to two wellfields consisting of 234 new wells drilled into the aquifer. A river indeed: it could carry 2 million tons of water a day, each wellfield contributing some 350 million cubic meters of water to the flow each year.

At the other end, on the coast, Ahmed and his co-workers were invited to watch the first few liters of water as they poured out of the pipe into a ceremonial tank. Libyan television captured their grinning faces. Sometimes propaganda is also truth.

Stage Two was to construct an even bigger wellfield to tap into the western aquifer at Marzüg. This one would consist of 484 wells and two 4-meter pipelines to take the water north. One of these lines would carry up to 700 million cubic meters a year to the coast and the farms of Gefara, mostly to stimulate growth there, but also to redress the environmental balance, restoring water tables and improving water quality. The other line was to transfer 175 million cubic meters a year to the communities along the northwestern mountain range.

Dong Ah brought both these stages off on time and on budget: $6.4 billion.

Three more stages were planned before the project would be considered finished. Stage Three has already developed an additional wellfield in the Kufrah oasis region, connected to Stage One's pipeline, adding another 560 million cubic meters a year to its capacity. Stage Four would connect the eastern branches to the western branches and take a further 350 million cubic meters to the Gefara plain. The final stage, Stage Five, would extend the eastern branch of Stage One to the city of Tobruk.

There have been some grumblings from all three of Libya's desert neighbors about what the colonel is up to. The presumption of most hydrologists is that the aquifers being mined spill across national boundaries, if these boundaries are extended vertically downward. But there is no proof: only if the water tables in Egypt, Chad, and Algeria begin to sink, and their aquifers become depleted, will evidence be to hand, and then it will be too late. So far, pumping tests and simulation models show no lateral flow from other countries, but all hydrologists acknowledge that it is too early to tell. Official Libyan policy is that groundwater aquifers, unlike naturally running surface water, should be considered "as any other natural resources of vertical utility," such as oil or minerals. And, as al-Ghariani pointed out in his paper, nations cannot really consider, at this stage of international law, whether "migrating resources" such as aquifers harm other nations using the same resource. Otherwise, he says, "the world would be overwhelmed by geopolitical conflicts that may arise among cohabitant nations as a result of induced pollution, whether due to circulation, geomorphological changes and migratory herds, birds, or humans." In any case, Algeria has its own vast sub-Saharan aquifer, under the plateau that slopes upward to the Ahaggar Mountains. Is that "theirs"? And if not, why not?

A few years ago there was a brief flurry of paranoia among Western security buffs that this $34 billion wasn't really being spent on water tunnels at all, but that the "pipelines" would be used instead to hide and transport troops. Like many other "intelligence analyses," this conclusion was neither intelligent nor an analysis; left unstated, for obvious reasons, was the question why Qaddafi would want to send an army haring off into the middle of the Sahara. Later a more sophisticated version of this notion surfaced: some of the subsurface "reservoirs" along the coast were not really water distribution plants at all, but deep cover for chemical warfare factories, built with Iraqi assistance. Adduced as evidence was an Iraqi passport that had been "clearly seen" in a government building in Tripoli. Despite the paranoia, of course, it might even be true, but meanwhile the water is flowing, and when the Libyans are asked what they will do when the water runs out, they speak of the American experience. The Americans, they point out, are mining their own aquifers in an unsustainable way. They've known for years that the Ogallala Aquifer under the High Plains states is being seriously overdrafted. What will they do when their water runs out?

It's a good question, and one that needs a little backing up to answer properly.

The Ogallala Aquifer lies deep in the shale and gravel beneath some 580,000 square kilometers of the Great Plains, particularly the portion known as the High Plains, ranging from West Texas, Oklahoma, and New Mexico in the south, through Kansas, part of Colorado, and Nebraska, and tailing off in South Dakota. The plains have always been a place of extremes — bitterly cold in winter with great driving blizzards, and torrid in summer. The land is difficult and treacherous to farm, for while the High Plains were once grasslands and not part of the Great American Desert proper, they had poor soils and only meager and erratic rainfall.

When the Spanish romantic and explorer Francisco Vásquez de Coronado passed through in the mid-sixteenth century, the plains were still grasslands, home to millions of bison, wolves, and grizzly bears. The Great Plains cultures of the Cheyenne, Sioux, Comanche, and Apache came into being only after the Spanish brought European horses from the Southwest in the 1700s. The first permanent white settlers brought cattle, which spelled the end of this nomadic native culture. For a few short decades, the flowering of the mythical Wild West and the Lonesome Cowboy, there were great cattle drives from West Texas to the railheads in Kansas, but the ecological consequences were not nearly so romantic: overgrazing, soil depletion, the beginning of desertification marked by invasions of mesquite and arid-land weeds, and prolonged droughts. When they went broke, the cattle barons made their way to California or back east. Their ranches were broken up, and only a few hardscrabble farmers remained, stubborn, tenacious, and poor.

That period ended during the First World War, when a series of wet summers and a booming economy attracted more prosperous farmers. Within a few years, the plains were a sea of wheat. But a few years after that,

> *The crop has failed again, the wind and sun*
> *Dried out the stubble first, then one by one*
> *The strips of summerfallow, seered with heat,*
> *Crunched, like old fallen leaves, our lovely wheat,*
> *The garden is a dreary blighted waste,*
> *The very air is gritty to my taste. . . .*
> (the lament of a farmer's wife, Edna Jacques)[3]

The farmers had come to the High Plains from the high-rainfall areas of the East. They brought with them eastern techniques, the worst possible for farming in the West, including the shallow plowing of marginal land, the destruction of groundcover and windbreaks, and an ignorance of what wind would do to sandy soil when it had nothing to anchor it.

In the 1930s the rains failed. Or, rather, the rains became normal, for over the centuries they had often failed and the landscape had adapted. But the farmers could not adapt. When farm prices collapsed early in the decade and the drought persisted, the unanchored topsoil began to move. Blowing topsoil drifted across roads and railroad tracks, keeping the towns and cities bathed in dust and grit inside and out, causing yellow twilights at midday. The roads were impassable, with deep drifts of sand that built up until they covered the fences, choked out the few remaining shelterbelts and gardens, and reached the roofs of chicken houses. On May 12, 1934, the Associated Press reported huge clouds from the Great Plains Dust Bowl at 10,000 feet over the Atlantic, and amateur statisticians began calculating the amount of arable soil that had been removed in this storm alone, reaching more than 300 million tons before the exercise became pointless. The Black Blizzard, the name ironically foreshadowing the laments of Uzbek peasants a decade or two later, swept from the Rocky Mountain states to Washington, D.C., and deep into the thoughts of Congress.[4]

When the Dirty Thirties were over, a few stubborn farmers remained. But the weather continued poor, the droughts stubbornly extending themselves. The farmers survived by sinking boreholes deep into the earth. A windmill could keep a household going, a few head of cattle, and a few hectares. A major study of the area, the *Great Plains Report* of 1936, blamed poor land practices for the disaster. It recommended that the most vulnerable areas be taken out of cultivation and put into rangeland for livestock. The Civilian Conservation Corps, at the report's urging, planted millions of shelterbelt trees and shrubs to break up the scouring winds and help anchor the soil.

Then came the invention of the centrifugal pump, along with cheap power from Texas oil wells and the hydroelectric plants at Hoover Dam. And, suddenly, pumps that had brought up a few liters a minute were replaced by devices that brought up thousands, easily enough to irrigate 50 hectares of cropland. Hydrologists soon confirmed what the farmers already suspected: that underneath the plains was a massive aquifer, the size of one of the smaller Great Lakes, that had been lying there, steadily expanding and accumulating its precious water, unmolested, since the last ice age.

Farmers who once faced bankruptcy feared it no longer. The crops would not fail anymore; irrigation water would see to that. There was no thought then for how long the water would last. A Travelers Insurance study of the High Plains in 1958–59 concluded that the future for irrigators was bright; there was no mention of the fact that water mining must have a finite end. But the numbers speak for themselves: at the end of the Second World War there were only a few thousand irrigated hectares in the whole basin; by the 1980s there were more than 7 million. A federal study of the same area, commissioned with great fanfare and a budget of many millions of dollars, was undertaken in mid-1982. It introduced the first note of foreboding in the hitherto sunny vision of irrigation's future. In 1914, the report said, there were only a hundred irrigation wells in West Texas. By 1937 there were more than a thousand. When the report was written, there were 74,000. While understanding the need for drawing on the aquifer ("the dependence on ground water introduces an element of control by society over water resources and their use"), the report noted that the "drawdown" was excessive and that the aquifer as a whole was "in serious overdraft." Almost 14 million acre-feet of water a year were being taken out and not replaced — as much flow as the Colorado River in a good year.

The city of Tucson, Arizona, is a good example of aquifer depletion. Until recently it relied entirely on groundwater wells. As the depth of the wells increased from 150 to 450 meters, the Arizona Water Management Act demanded that the city return to zero overdraft by 2025, a clearly impossible task because the available water is shrinking and the population is expanding. Some water is being imported from the Colorado River through the Central Arizona Project, but not enough to compensate for the overdraft. The only other place the city can get water is from farmers, and city managers are busy buying up farmland and its accompanying water rights, taking more farmland out of production. Nevertheless, there are promising initiatives. Arizona began to regulate groundwater overdrafts in 1980. Tucson itself has had water police since 1989, the same year it required low-flow appliances such as water-saving toilets. To some extent the initiatives have paid off. Tucson's per capita water usage is among the lowest in the United States, about 375 liters per person per day. But the reduction is too little and the demand still too great.

So, when? When will the Ogallala give out? The 1982 report, grudgingly admitting the probability of catastrophe, predicted it wouldn't happen until around 2020, which seemed a decent interval away. More recent studies suggest that this projection was overly optimistic. The real answer is complicated by geology and politics. The aquifer is not uniformly thick, and the rates of drawdown vary; also, the aquifer as a whole is too dispersed and fractured to act as a single pool of water would. Overall, the amount of land irrigated by Ogallala water has dropped 30 percent since 1978, but not all states, or even regions within states, are facing the same degree of crisis: Texas has already decommissioned nearly 1 million hectares, one-third of its total irrigable land, and West Texas is almost in panic mode, but Nebraska, at the northern end of the aquifer, has a positive recharge rate and has not shown much concern. The varying levels of concern extend to local politics as well. Congressmen from the area want to raise the national consciousness on the issue while retaining control in local hands; local people prefer local responses to local changes in the water table, and generally prefer conservation to bans on irrigation.

A pattern of unusually dry weather since 1992 has accelerated the depletion rate in the aquifer, only temporarily derailed by El Niño in 1998. The average annual drop in the water table was about 15 centimeters throughout the 1980s and into 1991, but thereafter it began to accelerate. In 1994 it was 60 centimeters, in 1995 almost a meter. The manager of the Ogallala Underground Water Conservation District, Wayne Wyatt, was not optimistic. Drought increases irrigation and decreases aquifer recharge, he pointed out. "It's a pretty serious change," he told a reporter in 1998. "It's impossible to gauge exactly when the aquifer will run out, but the current lack of rain will hasten that day."

More likely the pumps will never suck air. The cost of raising the water the extra meters required will have driven many marginal farms out of existence and imperiled others. But there's another problem. Most of the farmers have already cut down the shelterbelts planted after the depression — they got in the way of the mobile irrigation sprinklers — and have been shallow-cultivating marginal soils again. It's a perfect recipe for another Dust Bowl when the water does run out.

Only the memory of the Dirty Thirties, and the way the Great Depression and the Dust Bowl contributed to the migrations of "Okies" to the West Coast, can explain the political profligacy, not to say stupidity, of the Ogallala states and their agricultural policies. With prudence and rigorous conservation, the aquifer could have been made to last for hundreds of years. State agricultures would be smaller in scale, no doubt, and less rich, but a sustainable or almost sustainable agriculture was always possible. It was the Dirty Thirties — and the consequent never-failing wellspring of subsidies from Congress — that blinded people to the inevitable, just as the Soviet planners had been blinded by their neat little rows of capitalist-competitive production statistics in the Aral Basin.

If there is not to be a large-scale collapse — if 300,000 people are not going to be destitute and a substantial American industry consigned to the trash can of history — something will have to be done. But what?

A report issued in 1989, *Forecasting by Analogy: Societal Responses to Regional Climatic Change*, edited by Michael Glantz, put it tactfully: "Policies developed in response to depletion of the aquifer that may be technically and economically feasible must also be politically and socially acceptable. For example, the large-scale interbasin water transfers to this region from the Great Lakes or some other river basin may be sound technical projects, but they face considerable social and political opposition." To put it mildly.

If there is no water in the aquifer, and six states and a dozen cities depend on it, where is the water to come from? The Colorado and its tributaries are already spoken for. There was some notion that "surplus" water from the Mississippi Gulf area could somehow be lifted a thousand meters and transported 1,500 kilometers to the High Plains, and it was seriously enough mooted that Texans were asked to consider it in a referendum. This mad notion would have taken the Mississippi water across four major rivers and 150 minor ones, and the energy required to lift it the necessary distance would have consumed the entire output of twelve medium-sized nuclear power plants. No one any longer believes that the untold billions this plan would cost to rescue half a million farmers and some minor-league cities was politically or even economically feasible, but that didn't stop the water planners from floating, if that's the right word, an even more harebrained scheme — to redirect Canada's Arctic-bound rivers southward into the Great Lakes, and, essentially, reengineer the entire North American continent. No one thinks that feasible either. Perhaps the Canadians could be bullied into allowing it — though it's doubtful — but who would pay the billions it would cost, and who would benefit?[5]

That the water of the Ogallala will, at some point, simply run out is a given. Like all mines, it will exhaust its resource. The only real answer is rigorous conservation and the virtual abandonment by the six plains states of agriculture.

The problem, though, is that the bankrupting of agriculture in the region will affect more than a few farmers, or even a few states. Irrigation has made this second-rate farmland one of the world's breadbaskets. A significant proportion of American grain exports come from the Ogallala states, and almost half of America's beef. If, as is widely expected, China's water crisis drives it to the world market for grain, and if that increasing demand occurs when either the Ogallala Aquifer is exhausted or the water becomes too expensive to pump, that could precipitate a global food crisis. No one — not Argentina, Canada, or Australia — would be able to produce enough to make up the shortfall. Of course, the future is unreadable and forecasting the future a game for the foolhardy, but it's easy to see that a calamity could quite possibly follow. It's the human and political — the geopolitical — consequences that constitute water's most potent crisis.

All over the West — all over America — water overdrafting continues, perhaps not on the scale and at the rate of the Ogallala, but fast enough. Long Island, for example, gets its water from a closed-basin aquifer that is rapidly being depleted and poisoned by industrial runoff. And the water laws of the United States, which seemed so sensible in pioneer times, now only make things worse.

In the 1980s, for example, the Canadian businessman Maurice Strong became embroiled in a water dispute in Colorado. He emerged from it with his honor intact and his business almost so, but for a time he was portrayed in the local press in the San Luis Valley as a blood- (or rather water-) sucking Canadian, intent on siphoning the valley's precious resource and exporting it to Canada, a bizarre reversal of the usual Canadian paranoia about Americans.

Strong, through a complicated series of stock swaps and deals, became a principal owner of the Baca Grande Ranch, one of the historic ranches of Colorado, dating from a grant by the king of Spain to Maria Luisa Cabeza Vaca (later anglicized to Baca), who had accompanied Coronado on his expeditions to what was then Spanish America. The ranch dominates the northwest end of the San Luis Valley, including the peak of the highest mountain, Mount Kit Carson, and a corner of the sand dunes, an extraordinarily picturesque piece of desert nestled at the base of the peaks which has been made into a national park.

Underneath the valley is a massive aquifer — massive enough that some have suggested it might rival the Ogallala itself in volume. It is smaller in area but deeper, possibly up to 9,000 meters deep, though the volume of water it contains, and its exact structure, is still unknown. There is thought to be a smaller aquifer overlying a much larger one, and early assessments of the amount of water underlying the valley have been revised radically upward. The U.S. government, in the form of the Bureau of Reclamation, wanted to export the valley's groundwater to ease the demand on the Colorado River. Strong's company, too, wanted to sell the water to the Californians.

The laws governing water use in the United States are a unique product of the early struggles over water rights. One basic principle that actively encourages waste is the "use it or lose it" rule, under which the owner of water rights will lose them if the water is not put to constructive use. Strong disagreed with his own company's proposals. He wanted to keep the water in the valley and suggested a variety of "use it" proposals, including irrigation and a local brewery, before he had a falling-out with his partners. He withdrew his cash, and shortly after that the water court, and subsequently the appeals court, rejected the company's application and it had to be liquidated at a loss of well over $20 million to the remaining partners.

"Mining" doesn't always just mean overpumping. The Florida Everglades, the only real tropical wetland in North America, was essentially mined not to use its water but to get rid of its water to create land for farming, suburban development, and apparently endless strip malls. The summer rains that flooded Lake Okeechobee once created natural rivers flowing southward, a steady flow of fresh water for the Everglades. As Florida's population expanded, developers saw little use for swamp, and cut hundreds of miles of canals to divert the water — some 6.4 billion liters annually — to the sea. A series of dikes regulates the water that flows into what remains of the Everglades, but much of the water that enters Everglades National Park is contaminated by chemicals sent downstream from farms; there has been a 90 percent drop in the bird population since the 1930s, a direct consequence of poisoned water.

And yet . . . the water supply system has dismally failed to keep pace with south Florida's growth. A system created to provide drinking water for 2 million people must now cope with 6 million, and will have to deal with 15 million by 2050.

The Army Corps of Engineers, the great dammers, are now back in the game, commissioned to execute a twenty-year $7.8 billion restoration plan, Al Gore's pet project and the largest in U.S. environmental history. Several of the dikes that hold back the water from the Everglades will be removed; thousands of hectares will be bought by the government to act as wetland reservoirs; and thousands of wells, more than 300 meters deep, will be drilled to store excess rainwater for the dry seasons. Whether this will work is anyone's guess. Whether the whole plan will even be allowed to work is moot. Court challenges galore are in the offing from all sides: developers and farmers (the plan is too big, too costly), even the Sierra Club and eminent ecologists (the plan is too picayune, or will only encourage development by making water easier to get).

Water mining is occurring in more places than the United States and Libya.

Saudi Arabia is not just marginal desert, like the American West, the Aral Basin, or South Africa's Karoo. The Rub al-Khali, the Empty Quarter, is the real thing, 640,000 square kilometers (bigger than France and the Low Countries) of hard-core desert, containing steep dunes more than 240 meters tall and reaching temperatures in the high 40s (over 115°F) in the summer sun. Yet the Saudis — and the Omanis to the east, on the Persian Gulf — are exporting wheat and fruit, of all things.

There is some surface water and a few streams around the pe-

riphery of the Arabian Peninsula, particularly close to the old
capitals of the Yemeni empire in the south. But even in Oman,
where the government has recruited Western engineers to reinvig-
orate an ancient system of qanats (locally called *falaj*) siphoning
water from a subterranean aquifer, most of the water comes from
groundwater.

What surface water exists is being meticulously harvested, much
of it from seasonal floods. There are now more than 200 dams
with a cumulative storage capacity of more than half a billion cubic meters. Saudi Arabia is the world's preeminent desalter of water; Saudi government figures show that desalination capacity will
reach 2.5 billion liters a day by century's end, more than enough to
supply the country's drinking needs.

The fossil aquifers that supply almost all the country's water lie
deep in the rock, 300 to 500 meters underground. The thousands
of wells drilled into the sedimentary rock have shown no signs of
exhaustion. But groundwater depletion has been averaging more
than 5 billion cubic meters a year, and at that rate the water will
run out altogether in less than fifty years. The Saudis are aware of
the hazards. Research on desalination goes on apace. And if they
have to, they can divert agricultural water back to the cities, and
the country can go back to importing food — if they can find any
to import.

Even more examples: Indonesia has so depleted its underground
aquifers that seawater has seeped 15 kilometers inland; although
the country has no money, pipelines to bring the needed water to
the cities are expected to cost more than $1 billion. In Mexico City,
depletion of the aquifers has caused a dramatic 4-meter collapse in
ground levels; the Mexicans have been forced to find new water
supplies 250 kilometers away and several hundred meters lower,
building an expensive new pipeline to pump it uphill to the city.
In Britain, the chalk caverns underlying the Home Counties have
been emptying of water, and the region is facing the prospect of
serious subsidences; water engineers have had some success in

artificially recharging the aquifers with water taken from the Thames, among other places, prudently scrubbed of its many unpleasant additives. And I recently came across a plaintive query on the Internet from the little town of Dubbo in New South Wales, Australia. "Hi all," the message read. "I work for local government in Dubbo, and as with many other places, we are having problems with depletion of our aquifer. Dubbo is a rural city in central west NSW with a population of approx. 38,000. With increasing urbanization and everyone wanting green lawns and open spaces, we are having real problems. . . . I probably should point out that Dubbo's summer temperatures often top the 45 degree Celsius mark. Hence my question: Who out there is implementing an aquifer recharge program and how are you going about it????" (In typical Internet fashion, four question marks were used to encourage speedy response.)

And, indeed, back came a reply from an employee of the U.S. Geological Survey. "Hi Mark," it said chattily, "sounds like you've got quite a problem on your hands." Bruce, the guy from the USGS, appended the following message: "Regarding artificial recharge, the American Society of Civil Engineers just published a Proceedings volume of a recent international conference on Artificial Recharge of Ground Water," and he provided the Civil Engineers' marketing phone number. "Drop me a line if you have any questions about this," he said.

It would never do to underestimate human ingenuity. The engineers are not yet without options.

10

The Reengineered River

*If you turn a river into a sewer,
you can turn it back into a river again*

I N 1990 a fish was landed in the Sieg, a small tributary of the Rhine opposite Bonn, in the state of North Rhine–Westphalia, one of the most densely populated areas of Europe's industrial heartland. The fish turned out to be unfit to eat, for laboratory analysis found trace levels of mercury and cadmium that exceeded European Commission guidelines for toxicity in food for human consumption. That wasn't so surprising: no one had been tempted to eat the creature. But this was a good news fish story anyway, for it was an adult Atlantic salmon, the first of its kind to be found in the Rhine system since 1958, the year when accumulated poisons finally overwhelmed the ecosystem, confirming the river's status as the most prestigious sewer on the planet.

Now, one fish had survived the long journey from the North Sea to the industrial interior. That fish was very good news indeed.

On October 31, 1986, only four years earlier, a fire had broken out in an electrical switching box in a riverside warehouse in Basel, Switzerland, nearly 500 kilometers upstream from the place where the Sieg enters the Rhine. It was no one's fault. A mechanical system had failed. But flammable material was stored nearby, and that was definitely someone's fault. The fire spread quickly, engulfing a couple of storage rooms before racing into the main warehouse itself. The heat was intense and began to blister the steel drums stacked against the wall. Then one of the drums exploded.

By the time Basel's firefighters got to the warehouse, which was owned by Sandoz, one of Switzerland's largest chemical manufacturers, there was little they could do about the fire itself. After hearing from the owners what the building contained, however, they backed off, donned their gas masks, and worked feverishly to build a catchment wall to contain the runoff, to prevent it pouring into the Rhine. Too little, too late: the catchment collapsed, and more than 30 tons of poisons poured into the river, an evil brew of herbicides, fungicides, pesticides, dyes, heavy metals, and 2 tons of mercury.

A few hours later the toxic stew reached Germany. Bertram Mueller, a hydrologist who had been working with the ICPR, the International Commission for the Protection of the Rhine, saw it pass. "I thought the river ran red, but I might have imagined it," he said. "The dyes, I suppose. Otherwise, it looked no different. It was the Rhine I had always known. But I knew that as I watched, its creatures were dying. It was the most terrible feeling. I was frozen, sickened. I couldn't even cry. It was the worst case of chemical contamination of a European river ever, and maybe of a river anywhere, ever."

More than a million fish, it was estimated later, died in the catastrophe. Livestock along the banks were poisoned as they drank from the river. All the way to Amsterdam, 800 kilometers from Basel, engineers shut down the intake valves for drinking-water systems: they could cope, barely, with the "accustomed" level of Rhine pollution, but this was impossible. The reservoirs of a hundred towns ran on empty for weeks.

"We believed," says Mueller, "that the river would remain biologically dead for a generation. It was a despairing thought." Yet even this ecological disaster, perversely, was leveraged into good news. Chernobyl was still a recent and terrifying memory. The people and the politicians were frightened into action. The Rhine cleanup, begun with little popular enthusiasm more than thirty years earlier, was galvanized. They would, at last, do what had to be done.

The Rhine is Europe's most important waterway, rising in the Reichenau above Lake Constance in Switzerland and flowing for 1,320 kilometers to the Waddenzee in the Netherlands, most of it through Germany. Its catchment area of 185,000 square kilometers takes in Germany, France, Holland, Austria, Luxembourg, Liechtenstein, Belgium, even Italy, and is home to more than 60 million people.

The Rhine was known from medieval times not just for its legends (the Lorelei, the Nibelungen treasure, and the rest) but for its relentless focus on commerce. The venality and avarice of the Rhine toll keepers were also legendary, and the riverboat skippers mounted many an abortive plot to get rid of them. In the 1830s, as Europe's industrial engine began to roar, the vanguard of capitalism had little time for legend. The Lorelei were just in the way, and the Bingen Loch rapids, picturesque though they might be, and the inspiration for a thesaurus full of metaphors, just a menace to navigation. Thus began the makeover of the Rhine that continues today.

The first task was to drain the river's alluvial plain, the flood plain that was the source of much of Germany's groundwater and — though this was ill understood — the Rhine's natural flood-control mechanism. The impulses to do so were not all venal, and wetlands are not altogether benign: they breed mosquitoes, among other creatures, and insect-borne diseases were commonplace.

Behind a network of protecting dikes, villages and then cities colonized the plains, or, where there were no cities, the farmers moved in, finding the alluvial soil fertile for whatever crops they desired. This same pattern was happening all over the planet among industrial nations: along the Yellow and the Yangtze in China, along the Missouri and the Red in the United States and Canada, along the fens and waterways of East Anglia, and in Australia's Murray-Darling Basin. No one yet understood that wetlands were not just "swamp," to be drained and tamed, but were integral to river systems themselves. Within a hundred years, 80 percent of the Rhine's alluvial basin had been cut off from the river. Instead of seepage, there were sewer pipes; instead of springs, industry; instead of marshes, pesticides. Groundwater tables began to drop. More and more wells went out of production, and increasing numbers of farms began to rely on "imported" water, piped in from the Rhine, among other places.

The second task was to straighten the river to make navigation easier. These "corrections" (a nice euphemism) shortened the Upper Rhine by 82 kilometers and the Lower Rhine by 23 kilometers. The river was made more comfortable for the skippers, but because it was straighter, it flowed faster, and because it flowed faster, it carried away more debris. The channel began to deepen, scoured down 5 meters in places, dragging the surrounding water table down with it. In most years, dikes held the river at bay, but when it flooded, the floods were more savage than they had been before.

The third task was to "use" the river more intensively. Barrages were created for hydroelectric schemes. The huge barrage at Iffezheim was the one that finally did in the Atlantic salmon. The salmon is a robust swimmer, but the fish ladders built there never worked properly, and the salmon were never able to pass.

It probably didn't matter very much. Germany's economic boom after the Second World War saw the creation of many new industries, the expansion of several towns, and the creation of new ones, many of which poured unregulated effluent into the river. Fish were the early warning system, a measure of water purity. In the 1930s there were fifty-two species of fish in the river. By 1975 there were twenty-nine, and most of them were on the verge of extinction. The river was suffocating; oxygen saturation had dimin-

ished from a norm of 90 percent to a mere 40. A few years later biologists officially declared the river dead. It was already a media
cliché to call the Rhine the "Sewer of Europe."

It's not that the problems were being ignored. There was, indeed, a
constant rumble of discontent. The healing power of pristine nature had always played a significant role in German philosophy,
and the Green movement was politically potent there earlier than
elsewhere in the industrial world. German hydrologists displayed
none of the willful ignorance about the Rhine that their counterparts and central planners in Russia exhibited about the Aral Basin. As early as 1950, the key Rhine countries — Germany, France,
Luxembourg, Switzerland, and the Netherlands — formed the International Commission for the Protection of the Rhine, with a
mandate to examine the problem and propose action to correct it.
Little was done. Despite the increasing worry, Europe lacked the
appetite for gloomy stories of degradation and pollution. A boom
was what they needed, an economic miracle. It did not matter that
it had costs.

By the 1970s the accumulated problems had finally trickled over
into public consciousness. Cancer rates along the Rhine Basin had
started to climb. There were ugly algae blooms in the North Sea
and along sluggish sections of the river as nitrogen pollution increased. And dire warnings were issued about eating the remaining fish.

The money tap turned, and billions were spent on sewage treatment and industrial effluent control. For a while, it didn't show.
Industrial expansion and population growth kept pace with the
improvements forced on the system. By 1986, however, progress
was being made.

Then came Sandoz, Chernobyl, and the election victories of the
Greens. In September 1987, a year after the disaster from Switzerland, the Rhine ministers issued the following declaration: "The
ecosystem of the Rhine must become a suitable habitat to allow
the return to this great European river of higher species which

used to be found here and which have since disappeared, such as the salmon." In 1991, prodded into action, the ICPR came up with its Ecological Master Plan for the Rhine, which declared as its targets "the restoration of the main stream as the backbone of the ecosystem; and the protection, preservation and improvement of reaches of ecological importance."

"We mustn't forget," Mueller said. "We mustn't ever forget that political will, in the service of an informed and aroused citizenry, can still move the world. Even now, facing ecological catastrophe, we must remember that."

"The rebirth of the Rhine . . . has to count as one of the great environmental success stories of the century," said Germany's environment minister Angela Merkel, in 1997, with considerable hyperbole but some justification.[1] Gottfried Schmidt, a biologist who had been working the Sieg tributary in the ICPR's Salmon 2000 program, pointed out that the Sieg's water, once among the most polluted in Europe, was almost drinkable without treatment. Dozens of dams and other obstacles had been removed, wetlands were being restored wherever possible, and thousands of salmon fingerlings were being released. Schmidt figured that by century's end, somewhere up to 20,000 adult salmon would have spawned in the Rhine system. This would be just a tiny fraction of the historical levels, but a self-sustaining population of any size had to be counted a success.[2]

The ICPR was not stopping there. Although PCBs have been banned in the Rhine area for years, there are still persistent traces in the river's sediment and in fish: a 1995 study of eels caught in four of the Rhine countries showed alarming concentrations. Nitrate pollution is still going up, despite best efforts, mostly from wastewater treatment plants, cars, and power plants, but most of all from farm fertilizer runoff. Phosphates have been banned from detergents for years, but phosphate levels remain too high, and the commission in 1997 mandated another 17 billion deutsche marks to impose stricter thresholds. Heavy metal concentrations in the river are now down but not out. Cadmium is still found. Lead, copper, and zinc are present in concentrations five times higher than the ICPR's target values. Mercury in locally caught barbel exceed allowable levels, and the fish are still unsafe to eat. Serious floods in 1993 and 1995 meant that more banks would have to be restored and more wetlands initiated.

They will be. That's what's important. That's progress.

The Danube, however, is another kettle of effluents. One day Czechoslovakia (now Slovakia) came along and moved the Danube, its frontier with Hungary. It closed the old bed and made a new one inside Czechoslovakia. Then it dammed that one up to make electricity for itself, thereby drying up the Hungarian side. It destroyed Hungary's ancient wetlands, closed out a few tributaries, and dropped the Hungarian water table. The Hungarians were displeased, to put it mildly. They took the issue to the International Court of Justice in The Hague — the first-ever international water rights dispute to be so mediated.

Better there than on the battlefield, for tensions were escalating rapidly on both sides. Ethnic and political quarrels dating back centuries complicated any desire for settlement.

Unfortunately, both sides lost.

Of course, it wasn't so simple. These things never are.

The quarrel began a long time ago, when the Czech and Slovak entities were ruled by the Holy Roman Empire and its successor, the Austro-Hungarian Empire, whose casual disregard for natural justice and equity caused sporadic uprisings, which were ruthlessly suppressed. In the Napoleonic period, Czech nationalists maneuvered to set up a new state whose borders would reach the Danube and connect it to other nearby Slav territories. At the Paris Peace Conference after the First World War, they succeeded: Bohemia was joined to Slovakia, scooping in Hungarian communities on the Slovak side of the river. It then became Czech policy to demand Hungarian lands on the other side of the river too: the

Czechs had moved beyond mere consolidation to wanting unilateral control of the Danube waterway. This they didn't get, and when they came under the domination of the Soviet system, nationalist desires were suppressed.

In the 1950s, prodded by the Soviet Union, the two vassal states came up with a plan to alter the shallow reach between Bratislava and Gyor, and to join the river to the Danube-Main-Rhine Canal, the trans-European waterway. Moscow was pushing huge amounts of freight through the region, and in Moscow the plan made perfect sense.

Nothing much came of it until the 1970s, by which time the plan had been considerably refocused. In 1976 Czechoslovakia and Hungary signed a Joint Agreed Plan, this one to "fix" the Danube good and proper. For years, central planners in the two countries had watched the Danube waters rushing wastefully by on their way to the Black Sea, ending in Europe's only inland river delta. No doubt influenced by the grandiose plans being mooted in Moscow by their political masters, the Russian Communists, who were still considering the notion of pulling all the Siberian rivers southward to the Central Asian deserts, the Czechs and Hungarians wanted a Grand Plan of their own. The Danube was all they had. An important additional motivation was energy. Both countries were dependent on coal and oil to generate electricity; new dams on the Danube could produce up to one-fifth of the energy needs of each country.

They undertook, therefore, to build the Gabcikovo-Nagymaros Barrage system, whose declared aim was to divert the Danube into a new canal that would generate electric power, become the new route for international navigation, and stimulate economic growth in the region. There would be a 60-kilometer reservoir, two diversion canals, and two power plants. That there would also be serious environmental consequences was simply denied. It was an article of faith, and no studies were undertaken. This was a government-approved project. How could there be adverse consequences?

The reservoir, at Dunakiliti, straddled the border. A 17-kilometer bypass canal would divert 90 percent of the river into Czechoslovakia, leaving only a trickle in the original bed. Another 100 kilometers downstream, the same thing would happen with the Hungarian dam. Construction was supposed to start in 1986, with a deadline of 1990 for completion.

But the political landscape was changing, complicated by a serious economic crisis in Hungary. As the Soviet grip loosened, old ethnic animosities flared up once more. The Hungarians, prodded by the nascent democracy movement and a growing ecological awareness, dragged their feet. A new study showed that the dam project would cause serious problems to drinking water supplies, natural resources, and plant and animal life. The Czechs disputed the study: a little tinkering and the problem would disappear, they maintained. Nevertheless, the Hungarians persuaded the Czechs to push the deadline another four years into the future, and then, after construction had actually begun in 1988, they changed their minds altogether. They unilaterally suspended construction and demanded an abrogation of the 1977 treaty.

The Czechs were furious. Their response, in 1993, was the notorious Variant C, whose main intent was to divert the river into Czechoslovakia regardless of what the Hungarians wanted. They would shift the whole thing, and if the Hungarians didn't like it, tough.

This ultimatum raised two interesting points. The national frontier was in the center of the Danube. If the river moved, did the frontier? The 1977 treaty was vague. And if the frontier didn't move, that meant the river flowed through Czech territory. Could Czechoslovakia then demand tariffs for Hungarian goods being transported up or down the river through what were now "Czech lands"?

The Hungarian sense of paranoia, always finely tuned, detected another peril. There is a large Magyar population inside Czechoslovakia, and if they ever tried to revolt or demand sovereignty, the

Czechs could easily suppress them by threatening to strangle Hungarian trade by shutting down the river.

In 1992 the European Community, alarmed at the prospect of yet another conflagration within Europe — it already had Cyprus and the Balkans to worry about — inserted itself into the issue. It proposed a trilateral commission of experts to settle the dispute, provided that both parties guaranteed ahead of time to accept the panel's findings and that, while the study was under way, neither side would "engage in any actions which would prejudice the panel's findings." This meant, in the Hungarian view, that Variant C would have to be suspended, since it was unilaterally conceived and outside the reach of the 1977 treaty. The Czechs disagreed. Work on Variant C had already begun, and therefore it could hardly prejudice the panel's findings. Stopping it now would mean losing more than 2,000 kilowatt-hours per year, at considerable cost to the economy. Variant C would therefore go ahead, and construction would begin in October.

The negotiators went home.

On May 25, 1992, the Hungarians declared the 1977 treaty void.

On October 24 the Czech earth movers shifted into action, and the damming of the Danube began: Variant C was under way.

A day later the Hungarian cabinet met in emergency session.

The day after that, European Commission officials boarded their jets once more and headed off to Prague and Budapest.

Neither the Hungarians nor the Czech government of Vaclav Havel had any stomach for military solutions, and they allowed themselves to be bullied back into joint session. Negotiations began on October 28.

Before the year ended, the London Protocol was initialed by both parties. Variant C would be put on hold, except for work that was necessary to prevent flooding and to provide for safe navigation and protection of the environment. The Czechs agreed to maintain 95 percent of flow in the old Danube bed, and agreed not to turn on the power plant at Gabcikovo. The European Commission would appoint a committee of experts to review the environmental risks, water economy, navigational requirements, and flood control, and would study the possibility of reversing Gabcikovo. Both countries agreed that the International Court of Justice at The Hague would arbitrate the case. The Justices were encouraged to consider "all legal, economic and environmental matters."

It was the first time an international environment case had ever been submitted to the Court. It surprised no one that the first such case concerned water.

The Turks sent observers, and so did the Syrians. The Egyptians kept a close watch, as did the Israelis and the Jordanians. There are, after all, water disputes all over the planet. Everyone waited.

Where the Danube — the Donau, the Duna, the Dunav, the Dunarea, the Dunay — passes Bratislava, it is several hundred meters wide and a muddy brown in color, no longer the *schönen, blauen Donau* of Johann Strauss's romantic elegy. It has already traveled a considerable distance, rising in the Black Forest Mountains of Germany and traversing Austria. From here it bisects Buda and Pest, turns south to Belgrade, skirts the U-trap of the Carpathian and Transylvanian Alps, and passes south of Bucharest to its delta at Sulina, where it empties into the Black Sea. The river's collection basin touches on eighteen European countries over an area of 817,000 square kilometers, a territory inhabited by 87 million people. Like the Rhine, the Danube is no longer pure river but a manufactured waterway. To improve navigation, the slow, meandering sections were long ago dredged, deepened, and straightened; numerous canals were dug; and hydropower plants, barrages, dams, reservoirs, navigation canals, and locks were constructed. Water was extracted for agriculture, city uses, and industry; in return, sewage was dumped into the waterway.

The Russian researcher Irina Zaretskaya, who works with Igor Shiklomanov, has tracked the Danube in detail. She is clearly not impressed, and her distaste for what has been done is obvious, even in the formal academic prose of her technical reports. Of the Danube's engineering works she writes: "These works [carelessly] changed cross sections, coast lines, slope, bottom and suspended sediment discharge, as well as water quality." Elsewhere she has written: "The quantitative and qualitative depletion of water resources in individual regions of the basin has resulted in a critical situation, especially during dry periods. An increasing water resource deficit in the region can become a brake on the economic development of the countries." Her figures show that, by the mid-1990s, 31 percent of the Danube was being extracted for human use, and she expected the figure to rise to 37 percent in a decade. There was still enough water, but not by much. Availability was already only 60 percent of the European norm. In the driest year on record, 1949, the available water dropped to 1,500 cubic meters a year per person — not dangerous, but disturbing, "reaching critical values," in Zaretskaya's words. And the population is now considerably larger.

Bratislava doesn't help much. The city rises from the Danube on a steep hillside, a setting that would be picturesque had not wholesale degradation been permitted during the Soviet period. When I first visited in 1970, a pall of pollution hung over the whole place, the chimneys of factories belching horrid brown smoke into the atmosphere, their effluent pouring untreated into the river below. Things are better now, but not by a lot. Bratislava is the capital city of the independent entity of Slovakia, having shed the Czechs in the 1993 referendum (and Havel's Czech Republic shedding at the same time the whole Danube contretemps, much to its relief). The factories are still there, their effluent only slightly reduced. And the city still pours its sewage, untreated and innocent of any control whatever, directly into the Danube. Whatever else the Slovaks do or don't do to the Hungarians, they have already poisoned their beaches and made swimming hazardous to human health.

Zaretskaya tracked the deteriorating quality of the water as well as its reduced flow. Dozens of cities and half a dozen countries allow huge amounts of insufficiently purified storm runoff, industrial wastes, and agricultural pesticides to enter the river virtually unmonitored. No one has ever added up the total or catalogued the horrors. What is known is that the Danube pours about 80 million tons of contaminated sediments into the Black Sea every average-water-flow year.

The Danube is also a dire demonstration of how complicated overlapping environmental stresses can become, and how one problem can cascade into a series of others. When in 1972 the Iron Gates Dam was completed on the borders of Romania and what is now Serbia, it accumulated the fine silica that had been flowing freely into the Black Sea. Without this silica, the single-celled algae called diatoms, which used it to protect themselves in glasslike coats, began to die off. The drop in diatoms coincided with a dramatic rise in nitrogen and phosphorus concentrations from agricultural fertilizers, which led to explosive growth in algae populations. But because the diatoms were now unable to survive, other algae, most notably the dinoflagellate "red tide" organisms, were liberated, and poisonous red tides began to creep along the Black Sea coast. At the same time (in the early 1980s), a jellyfish native to the Atlantic coast of North America was accidentally released into the Black Sea from a ballast tank. The jellyfish population exploded, eating virtually all of the zooplankton that fed on algae, further encouraging the red tides. Their blooms consumed all the oxygen in the shallows, "and soon the rotten-egg stench of hydrogen sulfide haunted the streets of Odessa. Carpets of dead fish — asphyxiated or poisoned — bobbed along the shores. . . . The shallows where vast beds of seagrass once breathed life into the waters are regularly fouled in a fetid soup laced with a microbe that thrives in such conditions: cholera."[3]

A study by the World Wide Fund for Nature (WWF) found that the Gabcikovo-Nagymaros dams would only intensify the problem that already existed. They would allow the buildup of poisoned muck at the dams and turn the reservoirs into holding tanks for industrial and human effluent.

Most of the groundwater in the region — and the largest single supply of drinking water — comes from an aquifer several hundred meters below the Danube, in deep gravel deposits. The daily water output is about 1 million cubic meters for Hungary and 2.3 million cubic meters for Slovakia. In the summer of 1993, the supply was reduced to two-thirds the normal level. There was another fear: as the flow is slowed in the old riverbed, larger deposits of polluted silt and bacteria will infiltrate the groundwater. In addition, thousands of hectares of forest flood plain, agricultural lands, and Danube countryside have been lost. With the closing of the old riverbed, water in certain branches of the Danube fell by 2 meters, other beds dried out completely, and the groundwater table fell 3 meters. In July 1994, Hungary began to pump water from the depleted flow of the old Danube, mostly to revive damaged wetlands. In December, demolition of the half-completed Nagymaros dam was under way.

In April 1995, Slovakia and Hungary signed an interim agreement regulating the amount of water Slovakia could divert to its dam. The International Court was nowhere close to delivering a judgment, but politics demanded at least a show of agreement. Both countries had applied to join the European Commission and wanted to be on their best behavior; they wanted to show the other members that they were capable of solving disputes in "a European manner." It was all a bit rich, perhaps, after millennia of European savagery and warfare, and only a decade since the dismal years of Soviet domination.

The agreement also, if only *en passant,* defined the rights of national minorities in the two countries. Ethnic Hungarians in Slovakia had been disturbed by a decree — later rescinded — that they must "Slovakize" their names and abolish bilingual signs; the Slovaks, for their part, pointed out snidely that Slavs in Hungary had long been assimilated. Both countries, snideness aside, agreed that ethnic disputes, too, would be settled in that newly nice European way.

On September 25, 1997, the International Court finally issued its judgment. Slovakia was told, in blunt terms, that stealing the river and risking a unique environment in the process was illegal and must be reversed. Hungary was told it had no right to abrogate the 1977 treaty, and would have to proceed with a project it no longer wanted. The Court's reasoning was opaque, but essentially its judgment was that treaties are an essential underpinning to international order. If states had the right to dismiss them unilaterally, consistent and coherent international relations would be impossible. What the Court didn't say was that the 1977 treaty had been signed while the two countries were still signatories to the Warsaw Pact — itself a treaty unilaterally abrogated by several parties. The Court, while insisting that the 1977 treaty remained in force, never actually had the gall to argue that the Warsaw Pact was still valid.

The Court demanded that the two governments negotiate an environmentally sound solution. If they couldn't come up with one, the dispute would go to the United Nations.

Well, thanks a lot.

That wasn't the end of the saga. Early in 1998 the Hungarians and the Slovaks got together and Hungary announced, to the fury of its own environmental movement, that though the Nagymaros dam had been demolished, the government planned to replace it with two somewhat smaller dams to serve the same purpose — in the center of the brand-new Danube Ipoly National Park conservation area. This seemed, on the face of it, in clear contradiction of the Court's ruling. The environmental movement threatened to take the case to the European Commission and derail Hungary's bid for accession.

Slovakia, for its part, continued to run the majority of Danube water through its power plant, also a clear violation of the Court's ruling.

Meanwhile, the WWF had come up with a compromise plan that would call for destroying all the dams while still permitting some hydropower on both sides. It would also protect the old Danube riverbed. This "flow-through" solution — increasing the velocity of the river by narrowing and raising the riverbed, and introducing artificial flumes while still encouraging overspill into regenerated wetlands — would do away with the accumulation of poisonous sediment, pushing the problem neatly downstream.

By 1999 both sides were thinking about it. And so it goes . . .

Béla Liptak, a Hungarian academic and indefatigable campaigner against the whole project, was disappointed in The Hague court's ruling, but took comfort where he could: "This first international environmental lawsuit is a precedent case: it [was] discussed in Kyoto and was one of the catalysts in the recent creation of a World Commission on Dams. So although painfully, slowly, and possibly too late to save the Danube, we are making progress."

PART III

The Politics of Water

11

The Middle East

If the water burden really is a zero-sum game, how do we get past the arithmetic?

I REMEMBER having a conversation about water with a member of the Kabbalah, the mystical Jewish sect whose notion of a universal life force has been borrowed and traduced by the New Agers in their restless search for something meaningful in life. We had come down from the hilltop town of Safad, as it's rendered in English, the Kabbal stronghold, where the body of the seer Rabbi Isaac Luria was buried. On the steep western slope is a wonderful little fourteenth-century synagogue, a holy place of the Kabbal, where the son of a friend had gone for his bar mitzvah. The great Rav's body lies in a simple grave lower down the slope, and at the bottom is a perennial spring, where first the men and then the women go to be cleansed.

This spring is very old. It has been running, as far as anyone knows, forever. It was here when the town was built, which was before Solomon built his temple. It was already old when Alexander the Great clattered through on his way to glory (the inhabitants had rolled rocks down the hillside against the passing army, but the spring hadn't been tampered with — hardly worth the effort), old when Jesus walked these hills, and ancient by the time the Prophet was bringing about the Great Awakening down in the deserts of the Arabian Peninsula. The locals have hewn a bath

Chazaud

100 km
0   60 miles

Beirut

Mediterranean Sea

LEBANON

Litani R.

Hasbani R.

Damascus

Mt. Hermon

Tyre

Huleh Valley

Rosh Pina

Golan

Safad

Heights

BENOT YA'AKOV
BRIDGE

Tiberias

Lake Kinneret
(Sea of Galilee)

Yarmuk R.

SYRIA

EAST GHOR
CANAL

WEST

Jordan R.

Zarqa R.

BANK

Tel Aviv

ADAM
BRIDGE

Amman

ALLENBY
BRIDGE

Jerusalem

Dead
Sea

JORDAN

Gaza

ISRAEL

Negev
Desert

Sinai

EGYPT

Eilat

Gulf of
Aqaba

LEBANON

Hasbani R.

Dan R.

Banyias R.

Dan

Banyias

SYRIA

Huleh
Valley

10 km
0   6 miles

from the rock, and the spring water is clear and cold as a Greenland glacier.

I liked the Kabbalists; their worship is joyous, full of song and rhythm, but their devoutness runs to prayer sessions of eight hours or more, and I was relieved to be able to get away. My friend Isaac from Safad drove me down to Lake Kinneret, aka the Sea of Galilee.

What an incongruous setting! Just down the coast, at Tiberias, the fishermen who became Jesus' disciples had been caught in a storm and had seen the Master striding across the lake, walking on water. Now, as I stared across the sea, two water-skiers zipped by; nearby was the Sea of Galilee Parasailing School, and you could learn to hang-glide over the place where Christian pilgrims come to venerate the Miracle of the Water. A little later we went to have lunch at the Saint Peter Restaurant in the Ron Beach Hotel, and I started to laugh.

"Better order fish," I said, but the Fisher of Men was outside the purview of the Kabbal, and Isaac just looked puzzled.

After a while we went back down to the shore.

"This is where we get our water," Isaac said. "More than half the country's water comes from here. Tel Aviv and Jerusalem could not exist without Lake Kinneret. This is where the Jordan waters are stored. It makes Israel possible. Without water, we are dead."

"Without water, we are all dead," I said.

"No. Yes. I mean . . . that's true, we're all dependent. But it's starker here. I mean politically, not personally. This little river, and this little sea, give us nearly two-thirds of all our water. Also about three-quarters of Jordan's water comes from this valley."

He showed me a clipping he'd torn from the local paper. Since I couldn't read Hebrew, he paraphrased. "They're worried," he said, "about the water in the lake. There are some signs of saline intrusion. I don't understand the engineering, or where the salt is coming from because we're a long way from the sea, but the water people are worried that if the level of the lake drops too much — and the drought combined with increasing demand is threatening to do just that — the hydraulic pressure will ease and saltwater will intrude."

He thought for a moment. "They're diverting the Jordan River above the lake, to take water to the Negev," he said. "The problem with water in Israel is that we're already using all of it. There is no more. And the population is growing."

"Zero-sum game," I said.

"Yes. If we get more water, which we'll have to, it will have to come from somewhere else. But where?"

The conversation turned to other things, and after a while we returned to Safad for the evening feast, a groaning-board banquet with music and song and dancing, and the notion of water slipped away. Later that night I heard a rumble, which I took to be thunder, but it could have been military jets. It could also, Isaac told me the next morning, have been artillery. Safad is only 15 kilometers from the Lebanese border, and if you stand at the lookout on the summit, you can see the blue hills of Lebanon and the Golan Heights, once Syrian territory. Israel patrols the Golan now, but the Syrians are still there. Everyone in Safad can feel their restlessness. There were Syrian troops in Safad itself not that long ago, when the war was on. One never forgets about security in Israel — or water.

The day before I had driven to the north end of the lake to see the Jordan River itself. It was not much, perhaps 100 meters across, and sluggish. I had wanted to see the intake pipes for the National Water Carrier, but water is a security matter in Israel and I had been forbidden. The Water Carrier takes Jordan water to the communities of the coast, including Tel Aviv, and through an intricate network of canals to the irrigation schemes of the arid south. The Israelis, famously, are making the desert bloom, and the Water Carrier is how they do it. No one knows how much water the carrier is capable of transporting; popular assumptions are that it could carry the full capacity of the Jordan River, but officialdom is not saying.

A few weeks after I left Safad, early in May 1998, I read a dispatch in a European newspaper. It was datelined Tel Aviv: "Early yesterday morning, the main water pipe serving the south of the country was severed, cutting off water to more than a million people as temperatures soared above 30 degrees [Celsius] in the hottest April in 30 years. Israeli police were not sure whether it was a politically motivated attack."

They never did say, then or later. No one in Israel forgets that two-thirds of the water Israel uses originates in territory it now controls through military conquest, in the Golan Heights and the West Bank. Al Fatah, among other groups, has been targeting Israeli water institutions for thirty years. I thought back to Isaac's notion: "We'll have to get more water. But where?"

That was still the critical question.

By the mid-1990s, Israel was overexploiting its water, drawing down its aquifers at beyond replenishment rates by about 15 percent a year (2.1 billion cubic meters, against a supply that ranges in good years from 1.95 to 1.6 billion cubic meters in drought years). Jordan was doing even worse: it was using 20 percent more water than it was receiving. The coastal aquifers in the region, especially in the critical Gaza area, were seriously overpumped, and seawater intrusions were becoming a major problem — and a major political problem, given that Jews were allowed to drill their wells deeper than Arabs or Palestinians. The already potent Palestinian grievances were being ratcheted up by the brutal politics of water: if your tap runs only a day or two a week, and the Jews' taps run all week, well, that's an easy grievance to nurture, isn't it?

By 2010, according to Israel's own figures, it would have a water deficit of 360 million cubic meters. Jordan's deficit would be closing in on 200 million, and the West Bank's on 140 million. Considering that the Jordan River in a good year yields only 1.4 billion cubic meters and is already overstretched, where, indeed, is the water to come from?

Zero-sum game: the peculiar and fragile ecology of the region

has always collided ferociously with its complicated, fractious, per-
ilous politics, and also with the population dynamics and urbanizing economies. The Middle East has always been the place where
water wars are most probable. Indeed, Israel did have a shooting
war with Syria over water, and it is now widely accepted that the
1967 Arab-Israeli war had its roots in water politics as much as it
did in national territorialism. Israel controls the Golan Heights
for its water as well as for reasons of military security. Of course,
there are potent psychological and political reasons for keeping
the Golan Heights, but Israel doesn't need the Heights to keep
snipers away; modern artillery doesn't require line of sight, and
modern "snipers" could easily be 20 kilometers into Syria. Similarly, Israel maintains a military presence in South Lebanon at
least partly because that's where the water is, not just because it is
home to terrorist bases. In fact, the boundaries of the state of Israel
are to some degree the result of water considerations.

The water resources of the region are simple to outline, if difficult
to control. The Jordan River, the key to the system, begins in three
headwater streams. The Hasbani River originates in Syria and
has at least part of its outflow in Lebanon. The Dan and the
Banyias rivers originate in the Golan Heights, occupied in 1967 by
Israel and annexed in 1981; both flow into the Jordan above Lake
Kinneret. The lower Jordan River is fed from springs and runoff
from the West Bank and Syrian and Jordanian waters, and by the
Yarmuk River, which rises in Syria, borders Jordan, Syria, and the
Golan Heights (and so Israel), closely parallels the Jordan for several hundred kilometers, and empties into the Jordan at Adam
Bridge. The Jordan Valley is thus an international drainage basin, a
naturally defined area that defies national politics.
    Still, only about 30 percent of the water in the region is surface
water, from rivers. Groundwater, from the Mountain, Eastern, and
Coastal aquifers, accounts for the rest. The Mountain Aquifer is
the most substantial of the three, and gives Israel almost one-quarter of its total water supply. It consists of several major drainage

basins: the Western, which falls almost entirely within the boundaries of Israel; the Northeastern, which is located inside the West
Bank; and the Cenomanian-Turonian, under the northern West
Bank. The Eastern Aquifer basin contains a number of smaller
aquifers, all located within the West Bank area; 90 percent of the
water in the area comes from wells drilled into these sources. The
Coastal Aquifer, of which the Gaza Strip Aquifer is a part, has been
continuously overpumped for many years, not only by the refugee
population in the area but also by Israeli settlers tapping into it
from outside Gaza itself. It is not hyperbole to say that the Gaza
water system is in crisis: the pumping is far above the recharge
rate, and there is already so much seawater in the wells that the water is mostly undrinkable.

Water was critical to the founders of Israel and to Zionist planners. A viable state needed substantial numbers of immigrants,
and immigrants needed water for living, for farming, and for industry. As early as 1875, Charles Warren, a researcher with the British Royal Society, declared that, with proper control of water, Palestine and the Negev could easily absorb 15 million people. Chaim
Weizmann had demanded as early as 1919, at the Paris Peace Conference that followed the First World War, that the boundaries of
any future Palestine should include the headwaters of the Jordan
(Mount Hermon) and the lower reaches of the Litani River, where
it turns sharply west to head for the Mediterranean just above the
city of Tyre. In a letter to the British prime minister, David Lloyd
George, Weizmann had asserted his desire to have control of the
"valley of the Litani, for a distance of about 25 miles above the
bend, and the western and southern slopes of Mount Hermon."
Control of the Litani, the Jordan, and the Yarmuk had been seen as
critical to the future state's security. David Ben-Gurion in 1948 demanded the same thing: "The most important rivers of the land of
Israel are the Jordan and the Yarmuk," he said. But he also demanded that the boundaries of Israel include the southern banks
of the Litani.

In the end, Israel didn't get the Litani. The League of Nations rejected the Zionist claims, and the Litani became part of Lebanon instead. It is a constant frustration: only a narrow ridge separates the Litani from the headwaters of the Jordan, and it has always tempted Israeli planners to correct nature's mistake in this regard. The western leg of the river has been in the occupied zone for the last few decades, but so far Israel has not exploited the waters to any degree.

From the beginning, the Arabs' plans to protect their water necessarily contradicted Zionist plans to divert it to immigrants. Rising political tension in the region and the lack of a solution acceptable to all parties exacerbated the situation.

There were many early studies of and plans for the water in the region. The first comprehensive water plan that the Zionists professed to like was commissioned by the U.S. Department of Agriculture. Its author, Walter Clay Loudermilk, called for the irrigation of the Jordan Valley, the diversion of the Jordan and Yarmuk rivers for hydroelectric power, the diversion of water from the north to the Negev Desert in the south, and the use of the Litani River in Lebanon. He also recommended a canal to join the Mediterranean Sea with the Dead Sea — the first sighting of that bird of rare plumage, the Dead-Med Canal (later to be distinguished from its even more popular counterpart, the Red-Dead Canal). All these waterworks were to be controlled by Jewish immigrants, a notion that went down well with the Zionists but doomed his ideas elsewhere in the region. Loudermilk was dismissive. Any Arab who didn't like the plan, he wrote, was to be deported to the Euphrates and Tigris rivers.

The subsequent Hays Plan of 1948 was based on Loudermilk's notions, from which it differed only in its softer rhetoric.

If you take the twisty mountain road from Safad down toward Rosh Pina and then north to the Benot Ya'akov Bridge, and if you go north from there on Secondary Road 918, you'll find yourself winding into the Huleh Valley. Here, ecology and ideology have more than once collided, a swirling nexus in the anxious water politics of the region.

A hundred years ago, and presumably for millennia before that, this area was a swamp. By the time of the Palestinian Mandate in the late 1940s, it was still called the Huleh Marshes, and it seemed a fit place for the sturdy agricultural settlers of Zionist legend. The swamp was good for nothing but breeding disease and mosquitoes; malaria was rife, and so was yellow fever. It was once Syrian territory, but it became part of Israel under the Partition Plan of 1948. A year later, Syrian troops were evacuated and the valley became part of the demilitarized zone. In the Syrian interpretation, this meant that nothing was supposed to happen there. To the new state of Israel, however, it was an opportunity. Israeli settlers would irrigate the Negev with water from the Jordan River and drain the marshes, and the kibbutzim, the self-sufficient agricultural collectives, would turn the valley into productive farmland. The Seven-Year Plan, published in 1953, adapted the Hays variant of Loudermilk's plan and confirmed the intention to divert the Jordan River south toward the Negev. The plan included a comprehensive water network that would cover the whole country, the first mooting of the National Water Carrier. The Huleh Marshes were an integral part of the system; the first settlers arrived there in 1953, the same year construction began on the Jordan diversion at the Benot Ya'akov Bridge.

A month later, after furious Syrian objections — as well as U.S. and other Western economic sanctions against Israel — construction was frozen.

The Arabs, for their part, adopted a notion put forward by an American engineer, M. E. Bunger, to construct a storage dam along the Yarmuk River at Makarin, where three valleys join together. Bunger was a political innocent, ignorant of the buzzsaw of Israeli politics. Indeed, he largely ignored Israel, and set about trying to solve Jordan's and Syria's water needs, at the same time trying to resolve, at least to some extent, the Palestinian refugee problem by increasing the productivity of available agricultural lands in the East Jordan Valley and parts of Syria. Work on this project actually began in 1953, but Israeli opposition exerted pressure on both the United Nations and the United States to withdraw support. A short while later, Jordan and Syria discovered, to their fury, that work had simply stopped.

The political tensions over Huleh and the demilitarized zone escalated rapidly. Neither side felt it could give in. To Israel, the rights to farm the land and use its water were bedrock values, as basic as any in the Zionist philosophy. To the Syrians, this was theft, pure and simple. The Americans intervened with one more water plan. This one was produced by a special envoy, Eric Johnston, who modeled his scheme on the Tennessee Valley Authority. Johnston's expressed desire was to be more inclusive. He wanted to find what he called "an equitable distribution for existing resources" and a way for the countries to cooperate. Naturally, his plan was promptly shredded by all sides. No one liked it — not Israel, not Jordan, not Syria, not even Lebanon. (In later years this wholesale condemnation was interpreted as proof of the plan's essential fairness, and Israel largely adopted it.)

Later in 1953, Israel scrapped the Seven-Year Plan in favor of the more grandiose Soviet-style Ten-Year Plan. Among other things, the revision resurrected the Jordan diversion scheme of the National Water Carrier project. Prudently, the diversion point was shifted to Eshrd Kinort, at the northwest corner of Lake Kinneret, safely out of Syrian reach. Also prudently, the diversion was carefully designed not to overdraw Israel's water allocation under the Johnston plan.

That second piece of prudence, if it was intended to soften Arab animosities, was a bust. In 1959 Syria took the case to the Arab League, then chaired by Colonel Nasser of Egypt, and the League approved a retaliatory measure: the construction of storage dams on several tributaries of the Jordan and Yarmuk rivers, the Banyias, and the Dan. This was all entirely legal — the work was well within Syrian territory — but it turned on the Israeli buzzsaw just the same. The Israelis, calculating that their quota from the Jordan would shrink by almost 35 percent, issued carefully measured threats. For a while, nothing much happened. Construction work was approved in 1961 and actually started in 1964, and that was enough for the Israelis. Shimon Peres, then deputy defense minister, had already declared that water issues would be a cause of war, and now Prime Minister Levi Eshkol was even more explicit: "[This] water is like the blood in our veins," he declared. If the Syrians persisted, it would be they, not Israel, who had declared war. In response, the Syrians deployed forces along the border, and several bloody skirmishes resulted. Construction on the dams continued, however, and after Israel threatened "massive retaliation" on several occasions without result, it finally sent in planes to bomb the diversion installations, destroying them without effort. This was two months before the outbreak of the 1967 war.

With its occupation of the West Bank, the Gaza Strip, and the Golan Heights, and the outright annexation of the Golan in 1981, Israel tightened its grip on regional water resources as far north as Mount Hermon, headwaters of the Jordan. After the 1982 invasion of South Lebanon, Israel extended its command even farther, to include part of the Litani.

Israel has promised to return the Golan Heights to Syria, on a contested timetable. Seen from Safad, it doesn't look like such a good idea. Whoever controls the Golan Heights controls the water, and control of the water is key.

For another fifteen or twenty years after the original settlers first tilled the soil, farmers continued to move into the former Huleh swamps, now safely under Israel's control — "safely" if you didn't count the occasional shelling from guerrilla bases in Lebanon, but most of the farmers were used to that. As it turned out, however, there was a much more insidious enemy at work.

As the wetlands drained, the groundwater tables began to drop. Small streams and springs dried up, and wells had to be extended downward almost every year. Some of the farmers gave up, but others began to irrigate more heavily, adding salts to the soil in hazardous quantities. To deal with the new salinity, they switched crops to more salt-resistant varieties, but the same thing happened again as the saline levels increased past the plants' ability to adapt. Then the ground started to collapse. As the underground aquifers drained away, the earth subsided into the cavities left behind. In some parts the collapses were up to 7 meters deep, and occasionally whole houses disappeared. Agronomists and hydrologists probed at the soil, but it didn't take a genius to see what had happened: the now depleted aquifers had been essential for the health of the whole ecosystem; now that the water was gone, the region was dying. But if it was easy to see the problem, it was difficult to see what could be done about it. The farm lobby in Israel has always been a potent political force. Part of the Zionist ideal had been self-sufficiency, and the kibbutzim had become part of the Israeli identity. How could you shut any of them down?

In the end, it was beyond politics. There was nothing anyone could do. The subsidences continued. In the mid-1980s dust storms began to be seen in the area, salty soil picked up by the swirling summer winds, carried far to the south. Leachates started to appear in the Jordan, and drifted down toward Lake Kinneret.

By the 1980s, in any case, the political dynamic had changed. Water scarcity and rapidly increasing urban populations dictated new priorities. No longer were the farmers to be given everything they wanted. It was better to use water to manufacture goods, which could then be sold for hard cash, and to buy food. In other words, industry gave more value to the available water than farming did, and the inevitable happened: the last farmers were pushed from the Huleh, and the hydrologists moved back in.

The Jordan diversions for the National Water Carrier have also had far-reaching and unintended ecological effects. In 1953 the Jordan had an average flow of 1.25 billion cubic meters at the Allenby Bridge, near the Dead Sea; it now records a flow fluctuating between 160 and 200 million cubic meters, an eighth of what it was. And the water in the national water system itself is not as pristine as it should be: it contains higher mineral concentrations than are considered safe in Europe or the United States.

The reduced flow of the Jordan — its conversion from a river into what is essentially a drainage ditch — has had its effect on the Dead Sea. Already the lowest surface area on the planet, about 300 meters below sea level, it has dropped more than 10 meters since the early 1900s, and the drop is accelerating. The Dead Sea is important not just for its minerals (potassium and bromine factories mar its landscape) or its tourism potential (its famous inability to let floating humans sink has drawn thousands of tourists a year), but because its very volume has an ameliorating effect on the microclimate.

In the summer of 1998, the rattle of machine-gun fire could be heard at several points along the highway to Amman, Jordan's capital. There were no human or animal casualties: herdsmen were shooting holes in a state water supply pipe to get water for their stock. On the roofs of the capital there were frequent shouting matches and occasional fights as arguments erupted over accusations of theft from housetop water tanks. There were reports of assaults on private water sellers, who had tripled the usual price for 5,000 liters of water to 20 Jordanian dinars, or about $30, in a country where the per capita annual income is $1,100. The wells in the oasis of Azraq, once the home of Lawrence of Arabia, have been dropping for years, and in this long hot summer some of them ran dry. Nevertheless, corrupt officials handed out dozens of new licenses for well-digging. Rationing was imposed: most citizens could get water from public sources only two days a week. Each household was to get 85 liters a day — one-third of the quota in Saudi Arabia, and one-quarter of that allocated to each Israeli household. Meanwhile, in the wealthy districts (where water was not yet rationed), there were private swimming pools. And in the capital's public parks, the lawns were kept alive year round with sprinklers. In all, 50 percent of pumped water is lost, either to outright theft or to leaky pipes, some of them dating back to the time of the Palestinian Mandate.

It's not hard to see why there is rationing. The Jordan Basin (the Jordan and Yarmuk rivers) is the country's sole source of running surface water. Rain-charged subsurface aquifers are the only other source of fresh water. The kingdom began pumping water from this handful of underground aquifers in 1989. Its renewable water resources account for 650 million cubic meters each year. It is currently using 990 million.

In 1994, with the signing of the Jordan-Israel Peace Accord, there was, for a change, a dollop of good news. Water issues took up a substantial part of the text of the treaty. Among other things, the two parties approved financing for a Unity Dam on the Yarmuk, and Israel agreed to divert water from the upper Jordan River to Jordan during the dry summer months. Entitlements to Jordan River water — still loosely based on the Johnston plan — were agreed on. Water flow to Jordan would increase; jointly the two countries would repair the Dead Sea by importing seawater from the Red Sea via the Red-Dead Canal, or from the Mediterranean via the Med-Dead Canal. Either would cost around $5 billion, and the ecological effects were unknown. But the drop to the Dead Sea would generate huge amounts of electricity, which could be used to desalinate water. The desalinated water would flow in pipelines from Israel to Amman — peace water. No timetable was affixed, and there was no sign of the water. But there was, however fragile and tentative, an agreement.

In the middle of 1998, Jordan's veteran water minister and peace negotiator, Mintler Haddadin, abruptly resigned. Accusations and counter-accusations about mismanagement and allegations of being "soft on the Israelis" were bandied about, though hardly anyone seemed to believe them. No one knew what his departure meant to the accord, either, although there were many forebodings

about its being a "catalyst for broader confrontations," as a diplo-
mat delicately put it.

Apart from Turkey, Lebanon is the only other country in the region that has plenty of water. The farmers are chronically short of
water, and there is rationing in the capital, Beirut, but this problem
has more to do with the collapse of state authority in Lebanon after the disastrous civil war between 1976 and 1990 than with natural supply.

The Litani is one of the country's major rivers, with a flow of
about 580 million cubic meters a year, smaller than the Jordan but
with better quality (salinity is at a purely notional 20 parts per million, compared with Lake Kinneret's 300). It rises in the central
part of the northern Bekaa Valley a short distance west of Baalbek,
flows southward between the Lebanon and Anti-Lebanon mountains, enters a 30-kilometer-long gorge at Qarun, and abruptly
turns west, emptying into the Mediterranean north of Tyre.

The Israelis have never really given up thinking of the Litani
as rightfully theirs. And since almost half the water used in Israel
is already captured, diverted, or preempted from its neighbors,
why stop there? After all, Moshe Dayan, Israel's defense minister
during the 1967 war and a notorious war hawk, said that Israel had
achieved "provisionally satisfying frontiers, with the exception of
those with Lebanon." It is difficult to know to what extent water
motivated Israel to invade Lebanon in 1978 and again in 1982, and
to maintain its security zone there. The reason given was its need
to control guerrilla and terrorist bases in Lebanon, but water undoubtedly played a significant role.

Whatever the truth, the Israelis now have access to the Litani —
for a while. There are conflicting reports, never commented on by
Israeli authorities, that they are already using water from the
Litani, diverting it southward, building subterranean pipes to take
it south, or planning some other scheme. An FAO report in the
mid-1970s declared, on thin evidence, that Israel had built underground pumps along the Lebanese border to suck water south-

ward. In 1984 the Lebanese government lodged a formal complaint to the United Nations about Israel's supposed diversions, not only of the Litani but also of two smaller rivers that conveniently fell into the security zone controlled by the Israeli military. Some Western researchers have declared that Israel has been diverting southward about 100 million cubic meters a year from its own Hasbani River, and also some, "in insignificant quantities," from the Litani. Israel has consistently denied all these allegations, and it is fair to point out that most of the more indignant accusations about Israeli water grabs in Lebanon have come from politicized Arab sources.[1]

At present, Israel controls all the aquifers in the West Bank — as good a reason as any for not handing the region "back" to whomever it once belonged. About one-fourth of the country's water comes from the West Bank, an ecological happenstance that induced Minister of Agriculture Refael Eitan to declare on Israeli radio in 1997 that the country would be in "mortal danger" if it lost control of the Mountain Aquifer.[2] In the mid-1990s there was still considerable confusion about how much water there was in these underground reservoirs; majority scientific opinion seems to maintain that, with equitable distribution, there would be enough water in the region for both the Israelis and the Palestinians, but this advice is hotly disputed as resource analysis gets snarled in politics.

After the Six-Day War, all water resources in the area were centralized under the control of the military. In 1982, however, the minister of defense, Ariel Sharon, handed responsibility over to the National Water Carrier, which was supposed to integrate the Mountain Aquifer with the rest of the country's water resources. Despite this civilian control, Israeli water law still imposes severe restrictions on water use, for example, on anyone wanting to drill a well. "It is permissible," the law says with bland bureaucratic precision, "to deny an applicant a permit, revoke or amend a license, without giving any explanation. The appropriate authorities may search and confiscate any water resources for which no permit exists, even if the owner has not been convicted." In 1998 the number of licenses issued for new wells for Jewish settlers far outnumbered those for Palestinians. Also, Palestinian wells could not exceed a depth of 140 meters, though Jewish wells could go down 800 meters. Similarly, more than half the Palestinian villages still had no running water, whereas all the Israeli settlements did, another increment in the yardstick of Palestinian grievances.

The basic fact of the region is simple enough: Israel could not exist without West Bank water, and the Palestinians are entirely dependent on it. The meager available water is critical for both sides.

In Gaza, the home of the discontented, birthplace of the PLO and the Intifada, water has always been a critical issue. Recent salination of the water supply has only intensified the problem.

Gaza is small but densely populated, only 40 kilometers by 10, but geopolitically important far beyond its size. It lies on the Mediterranean coast where Israel meets the Sinai, sharing an 11-kilometer boundary with Egypt and a 51-kilometer boundary with Israel. Fewer than 1 percent of the people in the enclave are Jewish settlers; almost all the rest are Sunni Muslim Arabs. About one-third of its people live in UN-run refugee camps.

The climate is temperate but dry. Two-thirds of what rain does fall is lost to evaporation; the rest recharges Gaza's only natural freshwater supply, the Gaza Aquifer. This sandstone sponge, only a few meters below the surface, is thought to store a renewable 42 million cubic meters a year. Wells and boreholes currently extract 45 million cubic meters, and overpumping lowers the water table about 15 centimeters a year. As the water table falls, salt water from the Mediterranean and other nearby saline aquifers is introduced. Seawater has been detected up to 1.5 kilometers inland, and continues to threaten the entire aquifer. In many parts the water is so saline that it is damaging soil and sharply limiting crop yields. Citrus, Gaza's main agricultural product, is highly salt-intolerant, and is suffering from declining quality and crop yields. In addition, uncontrolled use of pesticides and fertilizers has chemically contaminated the groundwater. The refugee camps, most of which have no proper sewage control, have made things worse. One-tenth of the population simply uses the sand dunes as latrines, and poorly maintained septic systems frequently overflow.

Palestinian control since the Gaza-Jericho Accord of August 1993 has been rocky, and the Palestinian Authority has had little luck in reducing the ecological degradation. Talks in 1998 between the PLO and Israel covered "national territory," terrorists, and land for security in great detail, but hardly mentioned water.

But here, too, water scarcity, or what Thomas (Tad) Homer-Dixon, director of the Peace and Conflict Studies Program at the University of Toronto, calls "demand-induced scarcity," has played a large part in the continuing violent conflict. Population growth is the principal culprit. The mass exodus of refugees to Gaza after partition in 1948 more than tripled its population, which is now grown to almost a million people. Despite high infant mortality, Gaza has one of the highest growth rates in the world, perhaps exceeding 5 percent a year, and, as a result, per capita water availability has decreased dramatically.

Israeli researchers Nir Becker and Naomi Zeitouni have argued for ending all water subsides to the region, thereby discouraging farmers, heavy users of water, and encouraging instead industrial users, who make more efficient use of each liter. "Such policy," they argue, "will ease the shortage of water that is created by policies and behavior that were aimed indirectly at increasing the use of water."[3]

When, in 1967, Israel declared all water resources to be state-owned, strict quotas were imposed on Palestinian water consumption. As a consequence, thousands of citrus trees were uprooted, cisterns demolished, and natural springs and existing wells blocked. By contrast, Israeli water consumption in Gaza is subsidized, encouraging overuse and misuse. In 1995 Gaza Palestinians paid $1.20 per cubic meter for water, while Israeli settlers paid 10 cents.

As in the West Bank, water problems can be solved by neither one side nor the other but only by cooperation, however improbable that may seem. In a zero-sum game, someone wins and someone loses — unless the rules are changed.

12

The Tigris-Euphrates System

Shoot an arrow of peace into the air, and
get a quiverful of suspicions and
paranoias in return

A T THE OTHER END of the Mideast region, stretching from the Fertile Crescent to the Anatolian Highlands of Turkey, water matters are, if anything, even more fractured, with a history almost as long — perhaps even longer.

The Sumerians came first, as far as anyone knows, as an urban civilization. They were the first to use the Tigris and the Euphrates rivers as lifelines, as a storage bank of energy, and therefore of power, and therefore of prestige, and therefore of history and art. The Sumerians invented irrigation, and the crescent from Baghdad to the Persian Gulf bloomed, and they became rich. Then came the Assyrians and the Babylonians, then the conquering Greeks and the short-lived Romans. Capital cities rose, then fell, rose again, decayed. The Ottomans followed, and Islam. Nowadays you have the likes of Assad in Syria, Saddam Hussein in Iraq, and a succession of unmemorable Turks following the splendid Atatürk. You have Kurdish rebels (or "Mountain Turks," as the Turkish authorities call them) up in the Anatolian highlands, Kurdish PKK guerrillas finding safe harbor in Syria, and Saddam's twin persecutions of his own Kurds and the Marsh Arabs in the south, through whose villages his tanks rolled on their way to Kuwait. It's hard not to feel that somehow the Fertile Crescent has gone steadily downhill in the millennia since Darius the Great exhibited his extrava-

gant management skills by spreading his intricate irrigation network into the Syrian desert.

Both the Tigris and the Euphrates rise in the moisture-rich and
verdant mountain valleys of eastern Anatolia in Turkey. The Tigris
flows southeast, winding through low mountain valleys and across
the rolling Turkish plains, flirts with Syria (it's briefly the border
between Syria and Turkey), and then heads southeast through
Iraq, passing Baghdad and the Central Marshes before joining the
Euphrates to become the Shatt-al-Arab, which flows into the Persian Gulf. The Euphrates takes a more circuitous route. It starts
nearer the Black Sea, makes a sweeping curve west and south, and
wanders through a series of Turkish lakes and reservoirs before
crossing into Syria and Lake Assad, the reservoir of Syria's Euphrates Dam. From there it too heads in a southeasterly direction,
crosses into Iraq, and traverses the parched desert south of Baghdad before flowing through the Hammar Marsh — or the former
Hammar Marsh, for it has now virtually disappeared, its wildlife
and human inhabitants along with it — and so into the Tigris not
far above the Iraqi town of Basra.

  The Euphrates and its tributaries are Syria's major water source,
on which rests the country's hopes for increasing food production to match its burgeoning population. Iraq, downstream from
both Turkey and Syria, is dependent on both rivers. There are a
few tributaries to the Tigris flowing in from Iran's Zagros Mountains, but they don't contribute anything substantial to the flow.
Nearly 85 percent of Iraq's population fill all their water needs
from the twin rivers. The claims and counterclaims of the three
countries are in many ways mutually incompatible, complicated
by ethnic conflicts and historical memories. There is a long-stand-
ing suspicion of the Turks' perceived desire for regional domination, harking back to the expansionist aims of the Byzantine emperors. Turkey, as the upstream country, maintains that it has
absolute sovereignty over any water originating in its territory:
"The water is as much ours as Iraq's oil is Iraq's" comes close to

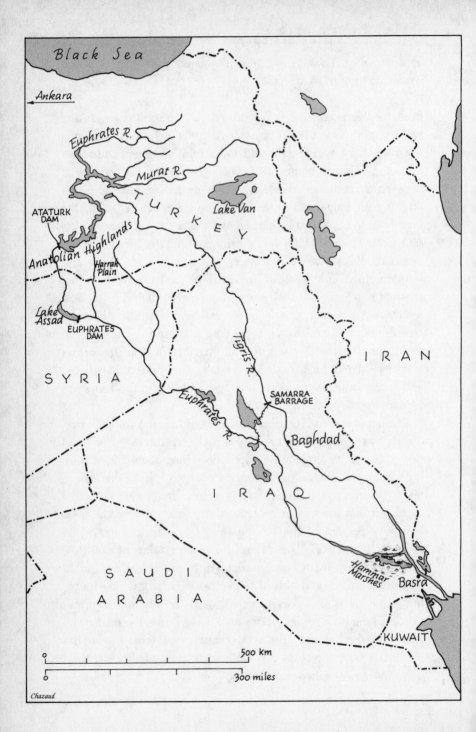

Black Sea

Ankara

Euphrates R.

Murat R.

T U R K E Y

Lake Van

ATATURK
DAM

Anatolian Highlands

Harrah
Plain

Lake
Assad

EUPHRATES
DAM

S Y R I A

Euphrates R.

Tigris R.

SAMARRA
BARRAGE

I R A N

Baghdad

I R A Q

Euphrates R.

SAUDI

ARABIA

Hammar
Marshes

Basra

KUWAIT

500 km

300 miles

Chazaud

outlining Turkish policy. Downstream Iraq, in contrast, argues its case on the rivers' "natural course," and on its "historical rights" to waters used by the people of southern Mesopotamia since the dawn of civilization six millennia ago, a prior-use doctrine that has been central to the struggle to develop international water law. Midstream Syria argues both ways: it cites the prior-use doctrine to Turkey but the sovereigntist one to Iraq, a classic case of speaking with forked tongue. This built-in rivalry is complicated by the presence of the Turkish Kurds, an ancient culture whose homeland straddles the headwaters of both rivers. The Kurds have been fighting a war of attrition for more than two decades, determined to win territorial independence and prepared to use guerrilla actions against Turkish hydro installations as a weapon of policy.

There is hardly a hint of a solution to these swirling controversies and disputes. Meanwhile, the amount of water available to all three countries continues to shrink on a per capita basis; Turkey's development plans will make things worse for Syria and Iraq, and Syrian plans will imperil Iraq's supply even further. Turkish government statements are hardly reassuring. In 1997 the secretary of state for water development, Kamran Inan, declared in a Turkish newspaper: "We give them [the Syrians and Iraqis] the water — they can't share it. Ninety percent of the Arab population is dependent on water originating outside [their borders] and therefore it is normal that the tension in the region will escalate." Such sentiments are not reassuring to hear in Baghdad and Damascus, especially from a non-Arab.[1] On another occasion Inan confided to a newspaper reporter that Turkey didn't feel "any international obligations" about the Tigris-Euphrates system, but softened his stance by referring to a "gentleman's agreement" not to harm neighbors. Nor is there any forum other than bilateral meetings in which these disputes can be resolved or even discussed. Turkey, whose political focus is primarily European, is nevertheless, at least for the moment, the dominant regional power both economically and militarily. As a consequence, Turkish water planners are paying little heed to the anxieties of downstream powers. The disputes in the region haven't come to war — yet.

Turkey is, by Middle East standards, water rich, but in truth it doesn't have much water to spare. Its total available supply is estimated at just over 193 cubic kilometers annually, which works out at slightly less than the total flow of, say, the Danube. In 1995, divided among the country's 60 million citizens, this amount yielded some 3,174 cubic meters per person per year, well above the water stress level. In practice, things aren't quite so rosy. For one thing, the population is increasing alarmingly; even a low projection has it at 78 million by 2025 (other projections put it closer to 90 million). Even the lower estimate would shrink per capita availability to somewhere between 2,100 and 2,400 cubic meters per year. Moreover, not all this water is easily reachable. Some of it is remote, and there are considerable difficulties inherent in damming and conveying it to where it will be needed. There are already shortages in the major cities, including Ankara, and along the Aegean coast, where tourism is gulping massive amounts of water that could otherwise be used for farming. In practice, therefore, most hydrologists assess the country's available water at 95 billion cubic meters annually, which would reduce the per capita availability to around 1,800 cubic meters — still adequate, but no longer rich, and around the same values for both Iraq and Syria, provided nothing changes.[2] But things are changing, and will change more.

Ilter Turan, a political scientist at Istanbul University, has been tracking Turkey's water needs. "It is anticipated," he wrote in 1993, "that consumption of water will rise rapidly in coming years. First . . . the government is pursuing a set of irrigation programs the completion of which will increase demands on the country's water resources. Second, the population is growing. . . . Third, the country is urbanizing rapidly. . . . Finally, the rapidly expanding industrial base creates new and additional water requirements." By the time per capita income rises to $10,000 a year, Turan forecasts, the country will need all the available water except the amount it contractually supplies to downstream riparian states.[3]

What will change the supply most is Turkey's grandiose plan for the southeastern Anatolia region of the Euphrates, a massive $32 billion collection of twenty-two dams, some of them huge, and an attendant network of irrigation canals, weirs, and barrages on a multitude of rivers, large and small, that would irrigate 1.7 million hectares of currently non-irrigated land — 1.1 million in the Euphrates Basin and 0.6 million in the Tigris Basin. The project will affect 74,000 square kilometers and change the way nearly 5 million people live, many of them Kurds. As the trial of Kurdish rebel Abdullah Ocalan got under way in 1999, Turkish officials admitted that one hoped-for result of the project's promise to farmers would be the quieting of separatist terrorism in the area. The centerpiece of the GAP (Southeast Anatolia Project, but known by its Turkish acronym) is the Atatürk Dam, near the town of Sanli Urfa on the Euphrates (Firat) River, the world's sixth- or ninth-largest dam, depending on whose statistics you believe. The dam was finished in 1990 and started to fill a year later. The water from its reservoir is carried to the Harran Plain in southern Turkey by the Sanli Urfa tunnel, a 26-kilometer behemoth that dwarfs even Colonel Qaddafi's Great Man-Made River (7.62 meters in diameter to the GMMR's 4 meters). A second string to this massive bow is the construction of no fewer than nineteen hydroelectric generating plants that would produce more than 27 billion kilowatt hours of electricity, adding 40 percent to Turkey's hydroelectric capacity, an amount equal to the country's entire power output in 1983. When completed — if completed, for money is short, and the World Bank has declined to write any checks without Iraqi and Syrian agreement — the project would involve the massive urbanization of Turkey's relatively poor southeastern provinces, and forecasts suggest that it would produce a 12 percent jump in national income. It would also, Turkish proponents point out, help regulate the alternating cycles of flood and famine, providing water in the lean years and preventing or minimizing flooding in wet years.

The entire project was originally scheduled to be completed by the late 1990s. That date was put off until 2005, and then, again, to 2010. Which was just as well: best estimates are that Syria will lose up to 40 percent of Euphrates water to the project. With its population growing at more than 3.5 percent a year, Syria was in danger of running out of water by early in the new century, even without the GAP. With it, the crisis will turn to a catastrophe.

Internal Syrian sources agree. Mikhail Wakil of the University of Aleppo was more measured but categorical: "Within the next two decades," he predicted in 1993, "Syria will experience an ever-increasing water deficit that will have serious effects on its social and economic development. . . . A water crisis is emerging throughout the Middle East, creating the potential for future conflicts." Sometime after 2010, Wakil's figures show, Syrian agriculture will need every drop of available water resources just to maintain the current per capita food production, leaving zero water for industry or domestic uses. And that's if the Euphrates supply does not diminish.[4]

Iraq, for its part, would lose somewhere between 80 and 90 percent of its Euphrates allotment. Iraq still has the Tigris, but that, too, will be affected by projected impoundments. Neither conservation nor more intensive use of smaller rivers would be able to make up the shortfall.

Even without the GAP, consumption targets for the three countries exceed the total supply by 18.3 billion cubic meters per year on the Euphrates system alone, and 5.8 billion for the Tigris.

Both the Syrians and the Iraqis claim ancestral rights to the water. The Syrians argue that the Euphrates is an international watercourse and that disputes involving it should be negotiated in an

international forum, something the Turks have adamantly opposed.

In 1987 Turkey agreed to provide a steady 500 cubic meters a
second at the border with Iraq, something it would never be able
to do if the GAP project were ever completed. Syria and Iraq initialed the agreement but continued to insist it wasn't nearly
enough; 700 cubic meters per second, a total of two-thirds of the
river's flow, was their real demand.

In January 1990 the Atatürk Dam was finally finished and
needed to be filled. The 1987 agreement to supply 500 cubic meters
per second notwithstanding, Turkey in effect stopped the river's
flow for a full month, to furious but ineffective protests from both
downstream countries. The Turks did issue advance warnings, and
pointed out, correctly if with overheavy irony, that it's impossible
to fill a dam if the full flow of water is allowed to pass it by. Also, by
increasing the downstream allocation somewhat before the stop-
page, they managed, with a little creative number crunching, to
prove to their own satisfaction that the "average flow" of 500 cubic
meters per second had in fact been maintained.

The low level of the Euphrates, aggravated by its poor quality
owing to agricultural pesticide runoff, chemical pollutants, and
heavy doses of salts, has on more than one occasion forced the Syrians to curtail drinking water and hydro generation for the major
cities of Damascus and Aleppo. Damascus is frequently without
water at night. The antiquated Syrian distribution system doesn't
help: as much as 35 percent of the water simply disappears somewhere along the cracked and leaking network of conduits.

Syria and Iraq are not much more cordial with each other than
they are with Turkey. They signed an agreement in 1990 sharing
the Euphrates — 52 percent for Iraq and 48 percent for Syria — but
not before they had almost gone to war on several occasions.

In 1974, when Turkey and Syria were both filling their new
dams, the flow to Iraq was reduced to one-quarter of the norm.

After a series of escalating threats from both sides, and Iraqi claims that more than 3 million farmers were suffering water shortages, Iraq moved troops to the Syrian border and threatened to bomb the Al Taqba Dam and its associated Assad Lake, constructed on the Euphrates with the aid of Russian engineers. In what was almost certainly a put-up job, the Iraqis the following day arrested a Syrian soldier, who had apparently been planting explosives in the holy city of Karbala. Syria denied everything, including lowering the river level, which was instead blamed on Turkey. Still, both sides mobilized and seemed on the brink of full-scale war before a team of Saudi and Russian negotiators helped cool things down. The dispute has never been settled, and a high level of fractiousness is maintained by Syria's repeated insistence that what it does with its water is none of Iraq's concern.

Complicating the problem were the shifting alliances in other regional conflicts: Syria against Iraq during the Iraq-Iran war; Syria and Turkey against Iraq in the Persian Gulf war. In the gulf war, Iraqi soldiers retreating from Kuwait set fire to dozens of oil wells and to Kuwait's entire desalination capacity. In return, the allied forces intentionally bombed and destroyed Baghdad's already antiquated water distribution system. In the early days of the war, moreover, there were attempts to persuade Turkey to threaten to turn off the tap against Iraq by diverting the Tigris, though this direct assault on hapless civilians fortunately remained uttered but not implemented. There also remained the perplexing problem of the Kurds, who have been waging a violent and bloody revolt against both Turkey and Iraq for national independence. The Turks have frequently blamed the Syrians for giving refuge to the PKK, the militantly Marxist Kurdish guerrilla movement that has killed more than 29,000 civilians in just over fifteen years and has promised to blow up the Atatürk Dam and the rest of the GAP as soon as it can. Turkey has more than once threatened to invoke its stranglehold over the region's water in its struggle with the PKK. In 1989 Turgut Özal, then prime minister and later president until his abrupt and unexpected death, said he would cut off the Euphrates entirely unless Syria expelled the PKK.[5] How he was going to do that was never explained (where would he put it?), but the threat worked anyway.

In late 1998 Turkish-Syrian relations were at an all-time low. Turkish newspapers were speaking openly of air strikes against Damascus, and leaves for soldiers stationed on the frontier were canceled. President Suleyman Demirel issued a terse statement to the Turkish news agency in which he said: "I am not only warning Syria, I am warning the world. This cannot continue." The Turkish chief of the general staff even declared, stridently and publicly, that there was a "state of undeclared war" with Syria. The situation was complicated by the PKK's public relations successes in western Europe — Italy allowed it to demonstrate openly in Rome — and Turkey's irritation at being kept out of the European Union. For its part, Syria has never scrupled to use Lebanese guerrillas as bargaining chips.

Iraq, too, has been sheltering Turkish Kurds while waging a bloody war of attrition against its own minority Kurds, including, notoriously, the use of nerve gases and other chemical weapons. Israel, to stir the pot, has been supplying arms to whichever Kurds will buy. After the gulf war there was a mass exodus of Iraqi Kurds to Turkey, further complicating Ankara's already difficult ethnic problem. Thousands of Kurds also fled to Germany as "guest workers," but increasing racial tensions there have been driving many of them home, and they are escalating their clamor for an independent Kurdistan.

The Marsh Arabs in southern Iraq have also been beneficiaries of Saddam Hussein's ethnic attentions. Iraq's "river project," the 560-kilometer artificial Saddam River which cuts between the Tigris and the Euphrates, starting near Baghdad and ending near Basra, was originally designed to "wash" salinated land and reclaim it for irrigation, an apparently benign ecological scheme. But a large area of additional land was deliberately flooded while the marshes were being drained, pushing aside the Marsh Arabs and making them easy victims of Saddam's pacification schemes. It was against the Marsh Arabs that Saddam first tried out the chemical weapons he later turned on the Kurds.

With this antagonism as background, Özal's surprise offer, while on a trip to the United States, of a "Peace Pipeline" to carry Turkish water to thirsty Middle East countries fell predictably on deaf ears. So did the more outré scheme to tow water from the Manavgat River, which otherwise flows harmlessly into the Mediterranean, to Israel or anyplace else that wants it, in so-called Medusa bags.

Still, the Peace Pipeline is nothing if not ambitious. There would be two branches, a total of 6,500 kilometers long, carrying more than 2 billion cubic meters of water a year from Turkey's Ceyhan and Seyhan rivers, which also flow into the Mediterranean, throughout the Middle East and Persian Gulf. The western branch would take water to Syria, Jordan, and western Saudi Arabia, eventually reaching Jidda and Mecca; the gulf branch would serve Kuwait, eastern Saudi Arabia, Bahrain, Qatar, the Emirates, and Oman. Both pipes would be buried 2 meters underground and would pierce mountain ranges in special tunnels. A study by an American consulting firm estimated that it would take fifteen years to build and cost around $20 billion in 1998 dollars, thus rivaling for cost and for degree of political improbability other Middle East water schemes: an aqueduct from the Euphrates in Iraq to Jordan, a canal from the Nile across northern Sinai to Gaza, and a diversion of water from Lebanon's Litani River via Israel to Palestine and Jordan.

None of the downstream countries particularly want to be beholden to Turkey for their water, no matter the protestations that the Turks would never turn off the tap. Everyone in the region has been obdurate in water negotiations with everyone else, and importing water is easily perceived to weaken other national aspirations. The Arabs don't want to rely on the Turks. The Israelis, ever mindful of the ease of sabotage of such a pipeline, don't want to be dependent on anyone, and don't want the notion of imported water to weaken their strongly held position that they are entitled to control all their domestic supply. Jordan, too, is fearful that accepting Turkish water will diminish its entitlement to local supplies. Kuwait most certainly doesn't want to depend on water coming from Iraq, nor does Jordan want to depend on the Syrians. Imported water is no one's first choice. The Turks, for their part, while occasionally speculating idly about the lucrative market the idea of exporting water opens up, protest their innocence. As Ilter Turan puts it, a touch indignantly, it's not as though Turkey doesn't need the water for itself. There are plans afoot for both the Seyhan and the Ceyhan rivers, and the idea "must be taken as a major gesture of goodwill" toward other countries in the region.

Indeed, shoot an arrow of peace into the air and get a quiverful of suspicions and paranoias in return. So much for water peace in old Mesopotamia.

13

The Nile

*With Egypt adding another million people
every nine months, demand is already in
critical conflict with supply. Another
zero-sum game?*

ONE EVENING not so long ago I was leaning on the railing of a creaking passenger vessel bound for Yemen, Mogadishu, and beyond, staring at the waters of the Rosetta Branch of the Nile Delta below. Off to the right was Alexandria, now a grim place, shabby and mean, but once an illustrious seat of learning, a beacon of civilization. There is nothing much left except a chamber of commerce that talks earnestly of glories restored, as though talking would make it true. Far off to the east was the other branch of the delta, at Darneita, near the mouth of the canal that cuts through to the Red Sea. There used to be more branches — the delta was once a maze of shifting channels — but they are gone, as much history as the Library of Alexandria. The shifting silt that made them used to come down from Ethiopia, but now it stops at Aswan High Dam or some other barrage, and the sea is reclaiming its own. I looked down at the water. Under a blood-red moon, the water was warm, sluggish, and black in the Egyptian night. For a brief moment a pair of dolphins splashed, a silvery glitter that was gone in an instant. I could hear the sound of a rough-hewn diesel, *pukkapukka,* across the harbor. There were low murmurs from the quay, where a group of Arab men in djellabas smoked and gossiped. The dolphins splashed again. They had swum in from Homer's wine-dark sea, here where it meets the Emerald River, flowing down from the Mountains of the Moon. Is there any more romantic spot on earth than this? Did not Mohammed, the Prophet of God himself, say that the Nile "comes out of the Garden of Paradise, and if you were to examine it when it comes out, you would find in it leaves of Paradise"?[1]

The Nile, the *nahal* or river valley in Old Semitic, is the longest river on the planet, wandering over 6,800 kilometers through 35 degrees of latitude. If you backtrack up the Nile from its delta, you'll pass El Qahir, that upstart city Cairo, only a few millennia old, and traverse the Eastern Desert, passing all the grand places of Egyptian history: Deir Mahwas, Qena, Qus, Luxor, Aswan, and Abu Simbel, Rameses II's monument to his own deification. From there you'll travel up the massive wound of the Great Rift to the African heartland. In Sudan you'll pass the places where the old empires of Kush and Meroë flourished; they governed Egypt at times, and in other periods Egypt governed them. Past Khartoum you'll have to choose, the White Nile or the Blue. The White and its many branches will take you through the Sudd Marshes, where the Nuer fishermen stand and wait, patient as storks, and on, south to the Great Lakes, to Jinja on Lake Victoria, to Burundi (where the Nile is really born), and to the Ruwenzori, the Mountains of the Moon. The Blue Nile will take you into Ethiopia, where legend says the Ark of the Covenant was hidden (and is still kept) after its disappearance from the Temple in Jerusalem — Ethiopia, whose culture goes back to the time of Solomon the Wise, king of the Jews. It was the annual drenching of the Ethiopian highlands that spills over into the Nile and the resulting deposit of silt in the delta that made possible the fecundity of Egyptian civilization.

When Nero ordered his centurions to find the source of the great river, they got no farther than the impenetrable marshes of the Sudd. Stanley, on his travels through Africa, recounts many legends of the Nile, most of them borrowed, in his magpie journalistic way, from early Arab manuscripts: "As for the Nile, it starts from the Mountains of Gumr. . . . Some say that word ought to be

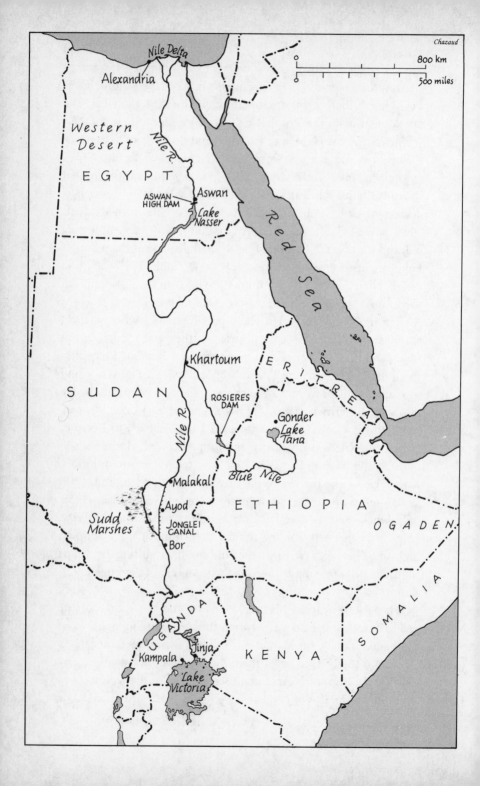

pronounced Kamar, which means the moon, but the traveler, Ti
Tarshi, says it was called by that name because the eye is dazzled by
the great brightness."[2] Veru, in his *Antiquities*, says Isis, Queen of
Many Names, had traveled down the Nile from Ethiopia to Egypt.
The Ethiopians, whenever they want to remind the Egyptians of
the source of their bounty, are fond of quoting this legend: more
recently it was trotted out by Mesfin Abebe, Ethiopia's minister of
natural resources and environmental protection, in a keynote address ("The Nile: Source of Regional Cooperation or Conflict?") in
Cairo to the Eighth World Congress on Water Resources in 1994.
The speech took as its theme the need for regional cooperation,
but Abebe spent a good deal of time being proprietary about the
Nile, reminding the assembled delegates that almost two-thirds of
the Lower Nile's water comes from the Blue Nile and from a tributary called the Atbara, both of which originate in Ethiopia.

Legends, lost in the suburban pages of history, but true enough.
The Nile is the Father of Rivers, the Seed of Civilization. Modern
engineers, at the urging of the pan-Arabist Gamal Abdel Nasser,
didn't do nearly as well as Idrisi of old. They put a stop to the Nile
Delta's infusions of silt by building the Aswan High Dam. No one
yet knows the full consequences.

The Nile is important — critical — to Egypt, which gets no usable
rain and has no other water but a few rapidly diminishing aquifers
under the desert. Only 2 percent of Egypt is not desert, and water
stress is rising every month. Egypt's 65 million people, climbing to
75 million by 2010, are entirely dependent on the river. The Nile
flows through eight other countries before reaching Egypt, the last
in line. According to a 1959 agreement with Sudan, the Egyptians
are entitled to 55 billion cubic meters of Nile water a year, while
Sudan gets 18.5 billion. In 1990, according to Sandra Postel, total
water availability in Egypt was 63.5 billion cubic meters, but in
1998 demand was already passing 68 billion and rising steadily.
USAID, the Agency for International Development, predicted in
the mid-1990s that Egypt would experience a 16 to 30 percent wa-

ter deficit by the end of the century. By 1998 that alarmist prediction had not yet come true, but the Egyptians were closing in on it. The targets for recycling water were not quite being met, despite more or less heroic efforts, and drought was reducing the Nile's overall flow.

Where is the extra water to come from? Egypt's water minister, Mohammed Abdel Hady Rady, maintained at the Eighth World Congress on Water Resources in Cairo in 1994 that Egypt was already efficient at water reuse: "Every drop is reused at least twice, and water efficiency is estimated at 75 percent," almost as high a rate as Israel's. Still, a study by Hussam Fahmy for the National Water Research Center suggested that Egypt's supply could be improved by capturing another 2.5 billion meters from the Nile by completing the Jonglei Canal with Sudan and by squeezing another 4 or 5 billion cubic meters from more intensive reuse of drainage water. Egypt has to some degree privatized agriculture, allowing farmers to sell directly to consumers instead of through the government, and is considering charges for irrigation water to offset the revenue losses. This trend is expected to improve efficiencies even further, since wasteful use of water now hits the individual farmer directly. In the Nile Delta, efforts are being made to reclaim land. New college graduates are given 2 hectares of land and a few hundred dollars a year to farm the land and feed themselves with tightly managed (and recycled) water supplies. The model? Israel's kibbutzim.

But even this development would simply bring supply up to meet demand, with no margin for error. In many other countries in the Nile Basin, water availability is on a collision course with irreducible demand. Tanzania, Burundi, Rwanda, Kenya, and Ethiopia are already critically short of water.

Egypt has said in the past that it is willing to go to war to prevent anyone upstream from tampering with its water flow, and there is no reason to suspect that its feelings have changed. In the mid-1980s it was on the point of ordering air attacks against Khartoum for perceived water threats, until cooler heads prevailed.

What has saved Egypt until now is that both Sudan and Ethiopia were too poor and wracked by civil war to contemplate any large-scale water diversions. It was why the settlement of Ethiopia's internal war and the destruction of its Derg tyrants raised such alarm in Egypt — and why the cynical Egyptians weren't particularly sorry that the country was settling back into its grim military fugue as the millennium neared its end.

Current tensions between Egypt, Ethiopia, and Sudan are in one way a continuation of a two-thousand-year-old struggle over who will control the water. Often in their history, Egyptian rulers have sought to unify the Nile Valley under their rule by conquering Sudan. In poor years, when the rains failed in Ethiopia and the annual flooding didn't come, it was common knowledge that human mischief was afoot. More than once an Egyptian sultan sent his ambassadors to the Ethiopian kings to plead with them not to obstruct the waters. The Scottish explorer James Bruce, who spent years in Ethiopia and at the Gonder court in the eighteenth century, recalled that the king had sent a letter to the Egyptian pasha in 1704 threatening to cut off the water. The current anxieties have only sharpened the potential for conflict. It's not just Egypt whose demand is increasing. Development, that Holy Grail pursued by all nations, will inevitably up the demand all along the river. The supply, however, cannot be changed.

The British, who controlled Egypt and Sudan on and off for a good part of the nineteenth and twentieth centuries, were the first to try to impose a basin-wide plan. To improve navigation, they sent sappers in to hack away at the series of natural dams and impediments that had built up over the years in Sudan's Sudd Marshes, impeding all shipping in the region. Like most of the developers who followed, the English were careless of the effects this would have on the marshes and the local climate, and on the local people, whose living was made from fishing. Before they could get very far, however, they had to leave, for Mohammed Ali, pasha of Egypt, sent mercenaries upriver to set up his own fiefdom at Khartoum. After Ali's death, control of northern Sudan became disorganized and corrupt. There were more and more European intrusions. The British, in particular, who were once again meddling in Egypt, returned to Sudan in a typically imperial mix of profit-taking, high-mindedness, and an eye for the main chance: control of the Nile. With Egypt falling into bankruptcy, matters became even more chaotic, and in 1881 the Sudanese rebelled in a fervent religious and political uprising called Mahdism, led by Mahdi Mohammed-Ahmed, who preached a version of Sufism. Among his victories was the massacre of the army of British general "Chinese" Gordon at Khartoum in 1885. The Mahdi ruled until 1898, when he was defeated and overthrown by an Anglo-Egyptian army under Lord Kitchener.

Shortly afterward, the sappers returned to the marshes in full force. By 1904 they had hacked a narrow path all the way through the marshes, and army engineers began drawing up plans to control and regulate the Nile's full flow. These plans were complicated by the fact that Britain dominated Egypt and Sudan but not Ethiopia. Although it went against the imperial grain, the British signed an agreement in 1902 with the Ethiopians that the Blue Nile would not be tampered with. The British also had to bully the French and the Italians into going along with their plans; both European powers had asserted rights in the basin. By the time they agreed, however, the Egyptian government was balking. It wasn't until 1929 that Britain was able to sponsor the Nile Water Agreement, which regulated the flow of the Nile and apportioned its use.

The Second World War postponed these useful endeavors. After it was over, the British, still in their own minds the political masters of the Nile Basin, commissioned what was to be the first thorough hydrological study of the Nile. Alas, once again it wasn't completed. The Ethiopians, restored to independence and jealous of their re-won prerogatives, declined to participate.

The study was finally released in 1958 as the *Report on the Nile Valley Plan*. It was in substance uncontroversial, but it did suggest a number of ways to increase the amount of water that reached

Egypt. The most critical was the construction of the Jonglei Canal from Bor to Malakal, finishing what the British had started seventy-five years earlier — to push a channel through the Sudd marshes and straighten out a "wasteful" U-turn in the course of the White Nile proper. The point now was not so much navigation. Where, after all, would the boats go? To Uganda? The point was pushing the waters of the Nile faster through the marshes. The canal would eliminate the enormous evaporation in the tropical sun during the water's sluggish passage through the marshes, freeing up what the report suggested would be 4 billion cubic meters a year for downstream users. Again, what this would do to the local inhabitants or to the ecosystem of the marshes was left unexplored. In the 1950s marshes didn't have ecosystems, or if they did, they were not perceived to benefit anyone. That climate is dependent on circulating moisture and evaporation patterns was ignored; climate didn't seem so easily tampered with in those innocent days.

The plan, put out with great fanfare in Khartoum, went nowhere. Prominent in its preamble was the notion that the Nile Basin must be treated, hydrologically speaking, as a single entity, and that the newly independent states would have to look beyond their own narrow interests and work together for everyone's benefit. This was a bit thick, coming as it did from a former colonial power, and only a few years after French and British paratroopers had dropped uninvited into Suez. Egypt, in any case, had lost interest in the Jonglei. It had developed plans of its own for building a dam at Aswan, and was busy playing the Russians off against the Americans to generate the necessary funds.

In Khartoum, the Egyptian plans for Aswan received a cool reception. A few years earlier, when the Sudanese had proposed a dam of their own on the Blue Nile at Rosieres to generate power for Khartoum, the Egyptians had been furious. Any reservoir above Khartoum would slow the Nile down, increase evaporation, and lose the Egyptians water. The Egyptians blustered and bullied, and the rhetoric became ever more extravagant, until the climate

was right for the Egyptian army to draw up battle plans. The Sudanese scheme for the Blue Nile came to naught, however, and the
crisis evaporated. Now it was the Egyptians' turn to build a dam —
and it would flood vast quantities of northern Sudan, force the resettlement of more than 50,000 hapless Nubians, and change life
on the banks of the Nile. Where was compensation to be found?

After a few years of simmering hostility, the two nations finally
got together in late 1958. Once they were settled around the table,
things looked a little easier, and in early 1959 they initialed an
agreement for "full utilization of the Nile waters." Sudan was pla-
cated by upping its allotment of Nile water from the 4 billion cubic
meters stipulated in the 1929 agreement to 18.5 billion cubic meters. It would now be allowed to build the Rosieres Dam, and was
encouraged to go ahead with the Jonglei Canal. In exchange, Egypt
would be allowed to proceed at Aswan without interference. The
treaty set up a joint committee to supervise all development projects that might affect the flow of the river.

There was one major drawback to this newfound cordiality: the
agreement divvied up the Nile's water without consulting Ethiopia, from which most of it came. Ethiopia, in retaliation, declared
it would reserve the right to use Nile water in whatever way it
chose, which once again sent Egypt's military planners to their
map books.

The hapless Nubians, up to 100,000 of them by now, were duly
resettled. Unfortunately, some had to be moved a second time
when construction started on the Rosieres Dam. The Sudanese
government promised compensation, but thirty years later no
money had been forthcoming, and many of the Nubians joined
the insurgents in the south.

The Aswan High Dam was finished in 1970, and Lake Nasser —
600 kilometers long and 50 kilometers wide in some places —
started to fill. The same year, Sudan and Egypt began joint construction of the Jonglei Canal, and were pushing it steadily southward when construction crews were driven out by guerrilla raids
by rebels of the Sudanese People's Liberation Army. They have

never returned, and nothing more has happened in the marshes since 1983. Sudan is still mired in a debilitating civil war. The two countries had spent more than $100 million of donor-country money and had nothing to show for it. Egypt needs the water the canal is supposed to yield, has paid real money for it, and is getting nothing. The go-it-alone growls from Cairo have been getting louder of late. Rumors — no more than rumors — say that another spasm of Egypt's military musculature occurred in 1994. The story is denied, of course, by all sides. But the following year there was an assassination attempt against President Hosni Mubarak of Egypt; Egyptian security officials didn't blame Sudan, exactly, but didn't exonerate it either. Border clashes between the two countries flared up again in the summer of 1998. Tensions remain high, and at this point the chances of Egypt's getting water from the Jonglei are more or less zero.

For the Egyptians, Ethiopia is an even trickier problem than Sudan. For one thing, it is farther away, and thus inherently less controllable. And at least the Sudanese are fellow Muslims. The Ethiopians are mostly Christians, when they're not being Marxists. They've always gone their own way, and still do. A month after Sudanese independence in 1956, and again during the Suez crisis, Ethiopia issued diplomatic notes that were highly undiplomatic in tone. Ethiopia, they declared, "reserved the right to utilize the waters of the Nile for the benefit of its peoples, whatever might be the measure of utilization of such waters sought by riparian states." In other words, buzz off. The Nile is ours to use as we see fit.

For a while it looked as though Ethiopia meant it. In the 1960s the government, at the personal urging of the Lion of Judah himself, the emperor Haile Selassie, hired the U.S. Interior Department, in the form of the Bureau of Reclamation, to develop a master plan for the Ethiopian Blue Nile and its tributaries. This was like honey to a bear: bureau engineers had spent half a century damming everything they could find in the United States, in legendary competition with their prime enemy, the Army Corps of

Engineers, and the notion of a tabula rasa for dam building — all those pristine, dam-free rivers flowing uselessly through to the sea! — must have been irresistible. They went ahead with sublime indifference to declared State Department policy, which was not to hassle the Egyptians in the aftermath of the Aswan debacle, in which the Soviets were thought to have scored precious diplomatic points among Third World countries. (The dam's evil side effects were not yet known, though Egypt's own engineers were already grumbling about it sotto voce.) In any case, the bureau's planners came up with a scheme that involved the building of twenty-nine irrigation and hydroelectric projects varying in size from modest to enormous. The bureau's own figures show that had all the schemes been implemented, the annual flooding of the Blue Nile — with its precious deposits of silt — would have been eliminated altogether, and the total flow would have been reduced by some 8.5 percent. Talk about provocative!

In any case, only one project was started, and even that was never finished: Egypt blocked the necessary financial approval of a loan by the African Development Bank. Then in 1974 Haile Selassie was overthrown by the Derg, a cabal of junior military officers with radical Marxist and dictatorial leanings, under the sinister influence of Mengistu Haile Mariam.

Mengistu brought his book-learned Marxism to bear on the Nile water problem. To develop an irrigated agriculture in the Nile Basin, he devised resettlement schemes along the tributaries of the Blue Nile and along the Baro-Sobat, a tributary of the White Nile. At the height of the most notorious human-induced famine in recent history, in 1984 the Derg pronounced that it would move more than 1 million people into these new settlements, and the first 15,000 were rounded up and bundled off to their new "homes." In the next year and a half, a million more were arbitrarily uprooted and moved, a piece of social engineering that caused immense suffering and thousands of deaths, and succeeded in irrigating enough land to feed about a hundred people.

It didn't work — so of course Mengistu did more of it. If you didn't have to confront what was actually happening on the ground, resettlement satisfied so many policy initiatives! It relieved population pressures in the cities, helped famine victims, settled down nomads, and provided land to the landless. So Mengistu announced that he would move ten times as many people as before.

The policy was an organizational and economic failure and a human disaster. Fortunately, Mengistu's control was never complete. The Eritrean guerrillas continued their war, and were soon joined by the Tigrai People's Liberation Front. In May 1991, with rebels only a few kilometers from the capital, Mengistu fled the country and took refuge in Zimbabwe. Eritrea gained (or regained, depending on your historical perspective) its hard-won independence, and a coalition of rebel groups under Meles Zeawi took control of a country with an empty treasury, a moribund economy, and a ruined agricultural sector. Despite this, they announced multiparty elections and a program of economic reform.

In a 1993 study reported in *Water International,* J. A. Dyer and his colleagues at Ottawa's Carleton University found significant potential for agricultural development in Ethiopia. True, parts of the country were arid — the Ogaden, for example — but "regions such as Keffer and Nyala have very rich land resources, capable of year-round production of food and cash crops." Other regions had potential, but would demand substantial irrigation. Irrigation would mean dams. And dams would alert the ever-vigilant Egyptians.

In 1998, however, fighting once again broke out between Addis Ababa and the Eritreans.

So far, nothing has been done in Ethiopia to damage Egypt's interests. But the Egyptians are always watchful.

In 1990, while Mengistu was still in charge, there were widespread reports that Israeli water engineers were examining Lake

Tana, the source of the Blue Nile, with a view to new irrigation schemes there. This notion was plausible: Israeli policy had always been to support the Ethiopians as a countervailing force to the dominant Muslim cultures of the region, and there were well-founded rumors that the Israelis were secretly helping the southern Sudan rebels against Khartoum.

In 1993 a general agreement was reached between the new Ethiopian government and Egypt. It was vague, but it did contain a clause in which each country agreed not to do anything to the Nile that might harm the other, and, perhaps more important, they agreed that "future water resource cooperation would be grounded in international law," without specifying what law, exactly.

But it is not so much what the Ethiopians have done, as Jan Hultin said in a survey for Leif Ohlsson's *Hydropolitics*, as what they might be doing. Former UN secretary general Boutros Boutros Ghali, when he was Egypt's foreign minister in 1990, pointed out that "the national security of Egypt, which is based on the water of the Nile, is in the hands of other African countries." At the 1994 conference in Cairo he was even more blunt: "The next war in our region will be over the waters of the Nile," he said, echoing a comment made by Anwar Sadat fifteen years earlier.

The potential for war is very real. But, with some political give and take, so is the prospect for regional cooperation. The best way for Egypt to "store" Nile water would be in massive reservoirs in the Ethiopian highlands, where evaporation rates would be far lower than they are at Lake Nasser, in the middle of the desert, which loses 2 meters every year to the sun, and in the 1979–1988 drought reduced the water level to a degree that threatened hydropower generation. Some studies have shown that enough water could be saved this way to quadruple Ethiopia's irrigated areas without affecting the downstream countries.[3] This solution would require a level of trust that none of the countries of the region have hitherto shown. But Egypt's leaders have not been strangers to vision, and they know the need is critical. No one really wants a war. If it comes, it will come only as a failure of political imagination.

In 1997 construction was supposed to start on an Egyptian scheme to deal with the fact that all the country's people live on a fraction of the land, a narrow cultivable strip along the Nile. The notion was to construct a brand-new "valley of the Nile" — a self-sustaining river that would flow through the Western Desert. A canal, called the New Valley Canal, would connect a series of oases, taking advantage of the excellent but waterless soil that exists in many parts of the country. A large number of people would be resettled there — not as the Ethiopians had tried to do, by arbitrary decree, but through land-holding incentives and subsidies. There were only two problems with the scheme: its estimated cost of $2 billion was money the Egyptians didn't have; and where would they find the water? They were already using — and reusing — their full allotment of Nile water, and every drop of it was needed elsewhere. The only way they could get more water was to renegotiate the agreement with Sudan. But Sudan said no — it, too, needed more water. Both Sudan and Ethiopia have fertile soil, easily irrigable, and in theory they could develop vigorous and productive agricultural industries based on irrigation. Of course, by doing so, they would be reducing Egypt's supply.

The only short-term solution available, apart from intensifying recycling and conservation, is to divert more water from farming — to turn more farmers into city dwellers, if the economy can absorb them. In the medium term, limiting population growth is essential; in the long term, regional cooperation. The same Cairo meeting to which Ethiopia's Minister Abebe delivered his keynote address failed to come up with any solutions, only pious platitudes. Abebe called for the appointment, at least, of a panel of experts to study such things as irrigation, hydropower generation, flood control, flow augmentation, fisheries, and the sound management of the environment. No such panel was appointed. In fact, despite Abebe's speech, Ethiopia was widely criticized at the meeting for obstructionism. It was noted, ironically, that increased silt loads in Sudan's dams could be blamed directly on deforestation in Ethiopia and its consequent erosion. Ethiopia flatly refused to discuss the issue, or any other issue it decided was "purely domestic."

At millennium's end, that was where the matter rested, with demand in critical conflict with supply.

14

The United States and Its Neighbors

*In the ménage à trois of Canada, Mexico,
and the United States, who is the seducer
and who the seducee?*

THE UNITED STATES is not running out of water, even of
clean water, and it's not going to go to war with either Mexico or Canada, despite the finely tuned paranoia in both
those worthy countries. This is not to say there aren't disagreements, or that the United States, like any imperial power in human
history, is not capable of taking what it needs, leaving rationaliza-
tions for those moppers-up of spilled policy, the historians.

The United States doesn't seem to realize how it is seen by interested outsiders: as energetic, creative, vulgar, acquisitive, and risible, with an astonishing record of pillaging the environment. But
in water matters the United States is both profligate and caring, rapacious and thrifty, and, when Americans turn their minds to it, of
awesome capability.

In the American West, water policy has amounted to what Marc
Reisner calls

a uniquely productive, creative vandalism. Agricultural paradises
were formed out of seas of sand and humps of rock. Sprawling cities
sprouted out of nowhere, grew at mad rates. . . . It was a spectacular
achievement and its most implacable critics have to acknowledge its
positive side. The economy was no doubt enriched. . . . Land that
had been dry-farmed and horribly abused was stabilized and saved
from the drought winds. . . . "Wasting" resources — rivers and aqui-

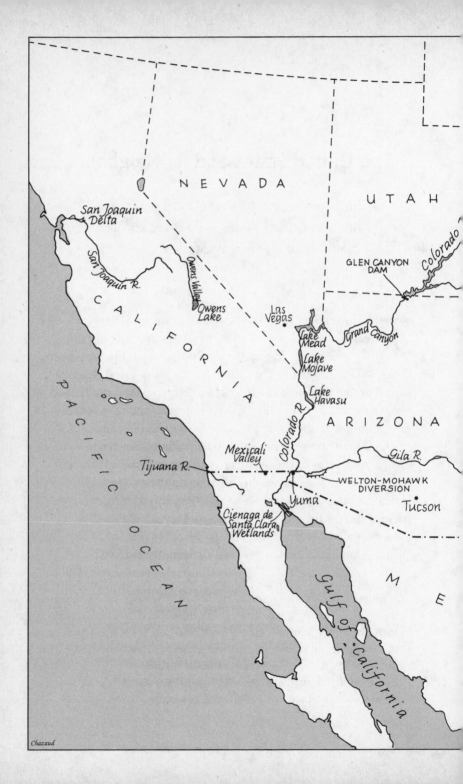

Chazaud

fers — were put to productive use. The cost of all this, however, was a
vandalization both of our natural heritage and our economic future,
and the reckoning has not even begun. Thus far, nature has paid the
highest price. . . . [But] who is going to pay to rescue the salt-poisoned land? To dredge trillions of tons of silt out of expiring reservoirs? To bring more water to whole regions, whole states, dependent on aquifers that have been recklessly mined?[1]

Well, Canada has water to spare, and Mexico virtually none. It
makes for an interesting ménage.

El Mayor, on the Colorado Delta in Mexico, is a scattered collection of ramshackle plywood and concrete block shanties on a couple of unpaved streets, rutted and pitted over time, dusty in the
dry years, and filled with mucky water when the rare rains come.
Children play, as they do everywhere in places like this, paddling in
the stagnant and dung-filled water, their playthings the shabby
castoffs of industrial civilization: rusting auto parts, scraps of old
tires, broken bicycles and baby carriages. El Mayor is too small to
appear on any map, and its "mayor's office" doesn't have a phone,
but it is still the principal town of the Cocopa people, who once
fished and hunted around here, innocent of the Yanqui scourge
to come.

Or maybe not so innocent. When the Spanish explorer Francisco Vásquez de Coronado first passed through in 1540, an aide
recorded the Cocopas' first response to seeing the European inter-
lopers: "They ran towards us, gesturing with fierce threats to
go back where we came from," he wrote. Would that they had
been more successful. The Colorado River that watered this place
has gone now, and the delta, rich fishing grounds for over two
thousand years, is a sad thing, desiccated, salt-poisoned, polluted,
nothing more than cracked mud flats, a rural slum. Aldo Leopold,
the American naturalist, canoed here in the 1920s and called it "a
milk and honey wilderness" filled with lagoons and dense forests.[2]
There were thousands of the Cocopa, then. Today there are fewer
than 1,500, living in straggling communities along the "Colorado"

and the little Hardy River, now no more than a drainage basin for farm runoff.

The water that does come down the Colorado from the United States is regulated in volume and quality, but has often been below the required volume and more saline than allowed. Some 90 percent of it has been drained away by U.S. cities and farms, to fill swimming pools in Los Angeles and generate electricity for the Vegas strip. By the time it reaches Mexico, there's very little left, and the towns and farmland of the border country take most of that, concrete channels distributing it to nearly 200,000 hectares in the Mexican desert. Mexico duplicates what the Americans have already done, creating farmland in an area where there should be none, growing crops in a region with less than 8 centimeters of rain a year.

By the time the water gets to the delta, and even when there are heavy rains upstream, a few steel culverts under a gravel road easily take care of what was once referred to as an "American Nile." In dry years, or when the Americans are recharging their dams, there is no water at all.[3]

The Cocopa fish, but in the Gulf of California now, where there are still fish, though many fewer than when the Colorado River flowed unchecked to the sea, bringing rich nutrients along with it. What it brings now is pollution.

American highhandedness with "Mexican" water goes back to the nineteenth century, to the years following the open warfare between the two nations. At the end of the century, skirmishing began again, this time over the Rio Grande, the border watercourse between Texas and Mexico from El Paso to the Gulf of Mexico. The Americans had been diverting water from the river to irrigate West Texas and New Mexico rangelands without agreement or permission, and the Mexicans filed a protest with Washington. The State Department, which handled border disputes, asked the U.S. attorney general, Judson Harmon, for a legal opinion.

The Rio Grande, said Harmon, rose in the territory of the

United States. The diversions were being made with American water. The United States had absolute sovereignty over its natural resources, of which water was one, no different in kind from gold or
iron. No international law existed compelling the United States to
share its water, and it would not allow itself to be bound by such a
law. The Harmon Doctrine, discredited now even in the land of its
birth, is still occasionally trotted out by recalcitrant countries trying to squeeze more water from transnational water systems at the
expense of downstream neighbors.

In the years that followed, the State Department devised more
flexible rules without ever actually repudiating Harmon. It
reached a tentative agreement with Mexico in 1906, and a more
thorough one in 1944 that incorporated not just the Rio Grande
but also the Tijuana and the crucial Colorado. That agreement was
carefully worded by State Department lawyers. Mexico was allocated specified quantities of water, a million and a half acre-feet
in the case of the Colorado. But there were interesting nuances.
The U.S. side viewed the allocations as unilateral — that is, they
were made by the United States as though there were no border
between the two countries. That meant that the Mexicans were
granted water, but not as a right.[4]

The Mexicans didn't argue. At least they had treaty rights to water in reasonable quantities.

What the treaty didn't mention, however — no one thought at the
time that it would be a problem — was the issue of water quality.
Water was water. You could irrigate with it, grow crops with it,
drink it. Sure, you could poison it, but no one anticipated that.

In the 1950s things began to change. Rapid industrialization of
the West, and the American desire to irrigate ever more unproduc-
tive land, began to change the turbidity of river water in the
boundary region. The Rio Grande has long since ceased being a
Texas legend. Since the 1970s it has been one of the most polluted
waterways in America, and was declared in 1993 America's most
endangered river, containing almost as much untreated sewage,

pesticides, and industrial waste as water, the unwanted and un-foreseen product of an unrelenting economic boom. In the No-gales region, water-related problems have been identified as the primary health concern for people on both sides of the border.

The problem with the Colorado started a long way from the border. Irrigation, as we have seen, tends to make runoff saltier, and so do storage dams. The Colorado's rush to the sea is first stopped at Glen Canyon Dam, near the Utah-Colorado border, by which time it has already received the agricultural runoff from the irrigated lands along the Green, White, and Yampa rivers. The river is stopped again at Lake Mead at Hoover Dam between Ne-vada and Arizona, again at Lake Mojave, and once again at Lake Havasu; at each stop the saline levels increase. Finally, it takes in the Gila River, the last tributary of the Colorado before it crosses the frontier into Mexico. There was irrigation along the Gila as early as the sixteenth century, when the Spaniards planted exten-sive gardens and feed grains in the valley. But the soil was clay, and the waterlogged soil was soon so salty that the crops died and the basin was abandoned. Then, in the 1940s, the United States built the Welton-Mohawk Irrigation and Drainage District of Arizona, which included a huge drain to carry away the salty residues. Outflow from this drain is dumped into the Colorado just before the border.[5]

By November 1961, Mexico had finally had enough. The govern-ment protested to the United States that "the water received is not suitable for agricultural uses, and agricultural production in the Mexicali Valley is being adversely affected." It insisted that the Americans were breaking international law by violating the 1944 treaty. This diplomatic language masked a growing popu-lar fury.

But instead of getting better, matters got worse. Two years after the protest, the Glen Canyon Dam was completed, and while it filled, the delivery of water to Mexico dropped sharply. The salin-ity of the Colorado rose to more than 1,500 parts per million from an annual average of less than half that, salty enough to poison most agricultural crops. The resultant Mexican uproar fell on deaf ears in the United States.

But by the middle of the decade, even the Americans had begun to be embarrassed, and in 1965 they undertook to do something about it. Mexico and the United States signed a five-year agreement promising "various measures," mostly unspecified, to deal with the salinity problem. Predictably, most of the state politicians and congressmen took the agreement for its results and, having agreed to do something, no longer felt the need to do anything. A few steps were taken on the local level, the sum total of which was to release some stored water when Mexico ran really low. This had the effect of reducing the salinity levels somewhat, but only down to around 1,245 from 1,500, still much too high. The long-simmering anger among Mexicans bubbled to the surface. The 1972 presidential campaign of Luis Echeverría was a Yankee-bashing affair based on the wretched state of Mexican agriculture. He won by a large majority. When the new president threatened to take the United States to the International Court of Justice at The Hague for treaty violation, President Nixon, eager to devise a sound way to do nothing, established a task force to study the problem, the classic politician's delaying tactic. He turned the issue over to the U.S.-Mexico International Boundary Waters Commission. Their job was to look at the Colorado Basin states (California, Arizona, Nevada, Colorado, Utah, and, through tributaries, Wyoming and New Mexico), as well as the Tijuana and Rio Grande rivers.

The task force hemmed and hawed, and came up with a variety of options in decreasing rank of political probability. In the end, it agreed that in all equity the United States should provide Mexico with decent water. But how? The options were simple but all of them expensive. The Americans, for example, could continue to pour salts into the river, in which case they would have to treat the water before it got to the Mexican border or substitute quality water from somewhere else, source unspecified. Or they could stop salting the river. The only way of making that happen was to take farmland out of production to eliminate the source of the highest salinity.

The two governments resumed desultory negotiations based on the task force's report. Probably nothing much would have been done had not 1973 been the year of the OPEC oil crisis and the consequent shock of seeing Americans in long lines waiting for gasoline. It came just at the time when seismic soundings showed Mexico to be, at least potentially, a major exporter of oil.

In 1974 the two countries announced that they had found a "permanent and definitive solution" to the salinity problem. Mexico was allotted the same amount of water as in the 1944 agreement — 1.5 million acre-feet of the Colorado River flow — and the United States agreed to meet specified standards of water quality, essentially the same standards as those offered American farmers.

Unerringly, Congress then opted for the most expensive of the available "scrubbing" options. The United States would build a series of drains and other engineering works on long stretches of the Colorado and — the linchpin of the scheme — build a massive desalination plant to process the water from the Welton-Mohawk Diversion in Arizona. This facility would desalt the water back to legal levels just before it entered Mexico. In addition, the United States would build at its own expense a drainage canal to carry the now massively saline effluent from the treatment plant directly to the Santa Clara slough in the Gulf of California, bypassing the Colorado entirely; and it would spend money to help Mexico rehabilitate and then "improve" the Mexicali Valley. The scheme was expensive, basin state congressmen acknowledged, but it had many advantages. They wouldn't have to take farmland out of production by buying out the farmers; they would minimize loss of water; they wouldn't have to discourage development and expansion in their own states; and the project would surely lead to important progress in desalination techniques. The total cost was estimated at somewhere around $1 billion. This way, as Marc Reisner put it, they would provide water to upstream irrigators at $3.50 per acre-foot, and then scrub the same water free of the resulting irrigation salts at $300 per acre-foot. The economics made no sense whatever.

And when all the salt has been removed, where does it go? Congress opted in the end for dumping it into the ocean through a diversion canal. The only other options were to evaporate it in large ponds, in which case mountains of raw salts would be likely to blow about on the winds, polluting land hundreds of kilometers away, or to use deep-well injection, an untried and untested technique that would force what is essentially brine into aquifers far underground, with unknown consequences. It seemed like a possibility, though, at a time when nuclear wastes were still being stored in "entirely safe" shafts deep underground.

The Colorado River Basin covers 632,000 square kilometers and provides water for 30 million people. This is the most watched, argued about, measured, and contested water in the United States. The river has an average flow of around 14 million acre-feet a year, which occasionally increases to 18 million in good years and decreases to around 12 million in bad. By treaty, Mexico must get 1.5 million of those acre-feet. By a complicated series of regulations, laws, and agreements, some of them of debatable legality, California has acquired the "right" to a minimum 4.4 million acre-feet, rain or drought. Evaporation from the river and its multiplicity of storage dams varies from year to year but averages around 2 million acre-feet. The upper basin, or "producer," states, which contribute to the river's flow, have an accumulated withdrawal allotment of around 3.6 million acre-feet. The Central Arizona Project, a 540-kilometer $4 billion canal, the largest and costliest water transfer system ever built, has the capacity to withdraw more than 2 million acre-feet, though it has been operating only at three-quarters capacity and is drawing just shy of 1.5 million acre-feet. In bad years that pretty well sucks the whole system dry. State-by-state allotments are 4.4 million acre-feet for California, 3,881,250 for Colorado, 2.8 million for Arizona, 1.725 million for Utah, 1.5

million for Mexico, 1.05 million for Wyoming, 843,750 for New Mexico, and 300,000 for Nevada.

The lower basin states — California, Arizona, and Nevada — have themselves been arguing about allotments for years. California has, by compact, used all the surplus water of the other two states, plus that of Wyoming, Utah, Colorado, and New Mexico, to water its cities and farms. But the rules are changing. The Western Water Policy Advisory Commission, in a 1998 report, speculated that of the ten American states that will grow fastest between now and 2025, five are in the Colorado Basin. The report uses typically extravagant western language: "It may be fair to say," it says, "that the major period of settlement in the West did not occur in 1850: it is just now taking place." Nevada, especially thirsty Las Vegas, already a city of more than a million people, wants its full allotment back from California, and then some.

As part of the reexamination that continues in all water-stressed states, California finally got around to redressing a problem of its own making: the damage it did to its own Owens Valley through the great water theft engineered by Los Angeles eighty-five years earlier, the subject of Roman Polanski's movie *Chinatown*. Culminating decades of bitter conflict, the Owens Valley and Los Angeles struck a deal in July 1998 that was intended to bring about an end to the massive dust storms that had plagued the valley for generations. The Los Angeles Department of Water and Power agreed to begin work at Owens Lake by 2001, and set a daring target date — eight years — by which the people of the Owens Lake area would once again be "breathing air that met federal health standards." Owens Lake is by far the single largest source of air particle pollution in the United States. For decades, people living around the Sierra Nevada have been exposed to the clouds of saline dust contaminated with toxic pesticide residues that swirl off the "lake" surface, little different in volume and substance from those that blow off the Aral Sea. The solution came when Los Angeles, for the first time in that thirsty city's history, surrendered water without a lengthy court fight. The lake will not be refilled, but the city agreed to keep enough water there to cover the bed with a thin film of moisture, while gradually treating the surface by planting salt-resistant vegetation and laying down gravel. This was the last major hurdle to repairing the damage Los Angeles inflicted when it opened its aqueduct and drained the valley of its mountain-fed water.

In late 1998, in an interesting case of life imitating art, water was about to flow from the water-rich but cash-poor farmers of the Imperial Valley to arid San Diego County, one of the most rapidly growing areas of southern California. The valley's water authority, the Imperial Irrigation District, agreed to sell 200,000 acre-feet of "surplus" Colorado River water to San Diego at $245 per acre-foot, more than twenty times the price it pays for the water in the first place. The deal began when the billionaire Bass brothers of Fort Worth, Texas, quietly assembled more than 200,000 hectares of Imperial Valley land in the expectation of acquiring water rights, which they could then sell to San Diego. The deal foundered when they discovered that the authority owned the rights, not the farmers. Inspired by this attempted chicanery (or, depending on your point of view, free marketism), the authority decided to play the game of marked-up prices itself. It first had to do a deal with the Metropolitan Water District, the Los Angeles behemoth, which owned the only aqueducts, but soon found a typically Californian solution: it persuaded the state legislature to pay off the Met through general tax revenues, thus giving the thirsty citizens of San Diego and the property owners of the Imperial Valley a massive subsidy, another neat way of benefiting the few while charging the many. Conservationists were left shaking their heads. Sure, San Diego's water supply was ensured, but if the precedent set off a rush of similar deals by other water-scarce California communities, the last barrier to unrestrained urban growth would be trashed.

The Imperial Valley farmers, according to the agreement, were to use the money they received to install conservation devices on their land. Conservationists pointed out gloomily that enforcement mechanisms were nowhere to be found.

Another hiccup in the process was the contention of the Coachella Water District that it had been unfairly deprived of water by Imperial — since 1934. Finally, in late 1998, in a speech to the seven states that depend on the Colorado, Interior Secretary Bruce Babbitt, who for a while took to referring to himself as the River Master, announced the settlement of the Imperial-Coachella dispute. Well, sort of: "I am not going away, I am continuing the pressure," he scolded the almost-partners. He threatened, among other things, to reduce the amount of water California takes from the Colorado unless the state learns to be more efficient, an assessment that was indignantly received by the profligate Californians. But the secretary was soothing: "I am very impressed," he said, "that Coachella and Imperial have at last discovered their fraternal bonds." He called it "a minor miracle." Indeed, in August 1999, negotiators for the Imperial Water District, the Metropolitan Water District, and the Coachella Valley Water District finally signed a comprehensive agreement. The impulse was not just Babbitt's threats but the realization that they were jointly facing a 37 percent increase in demand over the next two decades, with a far from assured supply. The declared long-term goal was to show the federal government and the other six Colorado River states that California was able to stop its internal feuding and learn to use its Colorado allocation frugally. In return, they hoped aloud that their newly self-defined status as conservators would persuade the federal government to "liberalize" (a nice word) its definition of what constituted a "surplus" in Colorado water, and thereby up California's allocation beyond its 4.4 million acre-feet. Reaction from the other states was unenthusiastic. Arizona, for one, declared itself unimpressed by these meager signs of Californian self-restraint.

No such luck in the northern part of the state. Not at all. In the San Joaquin Delta, the state's largest watershed, which provides water for 22 million Californians, the ecosystem was close to collapse. Babbitt had hoped for another "minor miracle" there, but it was not forthcoming. Agribusiness, environmentalists, and the urban water agencies could hardly agree even on an agenda for meetings, and Babbitt was reduced to calling them all "intransigent." The water agencies and the farmers wanted more water transfers; the environmentalists wanted conservation. Peter Gleick asserted halfway through the process that the state's Department of Water Resources "grossly underestimated potential water savings." Gary Bobker, another environmentalist, explained it this way: "There is an engineering logic to the idea that you can build a lot more flexibility into the system if you can bank more water in more pots. What is not proven is whether you can take more water out of the system and bank it without doing harm to the environment."[6] Delayed for what was said to be "more study" was the most divisive issue facing CalFed, a federal-state "superagency" that was supposed to come up with a thirty-year $10 billion plan for the area: whether to build a 42-mile canal to divert Sacramento River water around the delta and so more directly to southern California. The absence of this canal is an intense irritant to southern water districts, which see the stubborn north as awash in water it doesn't need.

As the century closed, an army of negotiators from dozens of agricultural districts were preparing for difficult negotiations with federal officials. They were all facing steeply increasing demand for water, and were lobbying vigorously against a congressional mandate that some 800,000 acre-feet of Central Valley water be set aside for "environmental mitigation."

"It's going to get extremely ugly," water lawyer Mark Atlas told the *Wall Street Journal* in December 1999. Atlas represented half a dozen water districts in the talks.

Ugly, indeed, is probable. The Bureau of Reclamation was mandated by the same act of Congress to recover all Central Valley water costs by 2030, but by December 1999 it had collected only about 6 percent of the capital expenditures and less than 2 percent of the public funds laid out as water subsidies. The only way it could get more was to increase prices steeply.

For some time Babbitt tried diligently to make sense of the maze of regulations, entitlements, laws, and agreements, without much result. In the end, he authorized a form of water banking that allows exchanges of Colorado water between lower basin states. In practice this scheme benefited mostly Arizona, which had been taking but not using its full allotment, prudently storing the surplus water in underground aquifers. The rules allow Arizona to bank the water in aquifers for Nevada. For Nevada to "buy" the water, the state dips it from the Central Arizona Project and issues a credit to Arizona. In return, an Arizona water user, say, the city of Tucson (which otherwise relies on rapidly depleting deep wells), pumps the same amount of water from the aquifer underlying Arizona, with Nevada paying all the costs.[7] Nevada, in turn, subsidizes its water to its users. Water traded in this way costs about one-third of what it does in Santa Barbara, California — which until recently used its "own" water, without joining the California distribution system, and saw its rates skyrocket when it finally signed up.

Of course, the system encourages ever greater use. Nevada can now get more water than its entitlement, and can attract more industry and ever more people — which means it will need even more water. Nevada had some warnings for California: "There are seven sovereign states on the river, not one sovereign state and six lesser partners," Patricia Mulroy, general manager of the Nevada Water Authority, told a *Los Angeles Times* reporter in August 1999. "California needs to remember that."

It is entirely possible, though, that the Colorado River will be in overdraft well before the scheme is implemented. There are still half a dozen major diversion projects on the drawing boards of seven states. If only half of them are built, the river will finally be overdrawn, and extra water will have to be found from somewhere. But where?

Mexico hasn't got any to spare, though its 1.5 million acre-feet of Colorado entitlement is vulnerable. Some northern California rivers are still running free to the ocean, but the long history of animosity between the north and the south of that state makes implausible any more diversions from that source. There are untapped rivers in Oregon and Washington, too. And if the provisions of the North American Free Trade Agreement are read in a certain way, if water can be certified as a commodity under the trading rules — well, then, Canada has lots of water, doesn't it?

Early in 1998 there was a sudden little ripple in the otherwise placid waters of Canadian politics. It was caused by a permit granted by the province of Ontario to a small company, based in the border town of Sault Ste. Marie, to export 600 million liters of Great Lakes water via cargo ship to Asia. The very notion was improbable. Water, though precious, is not yet valuable, and it is heavy. The company concerned, Nova Group, declined to say how it would profit. Where would these tanker ships go? You can't sail from the Great Lakes to Asia without traveling first down the St. Lawrence River, down the east coast of North America, and through the Panama Canal or around the Horn. What all that distance would cost in freight charges alone never came up. And who would buy the stuff? Who in Asia was that short of water? Nova didn't say. It hadn't yet looked for customers. It didn't have time: the mere notion brought the anti–free traders and the cultural nationalists rushing to the barricades to protect *our* water from *their* greed — "their," in this case, for once, not meaning the Americans. Except indirectly. Water exports, the Council of Canadians declared, was a moral issue.

Running through the argument were consistent errors. The notion that Canada has "two-thirds of the world's freshwater resources" became a commonplace in news stories, though it is entirely wrong (the real number is just shy of 6 percent of annual global runoff). There were even reports that said the Great Lakes themselves contained half the world's fresh water, which was not only wrong but ludicrous. Still, water is copious in Canada: by some measures, Canada already "exports" about 79,000 cubic meters of water per second as it flows into the sea or across the border, and the Nova Group proposal would have taken, each year, a volume equal to about ninety minutes' worth of the average annual outflow of the Great Lakes Basin. Put another way, 600 million liters is the amount of water that flows into the St. Lawrence River every minute. Lake Huron alone is the size of the entire Ogallala Aquifer.

"You'd think there wouldn't be any problem," said Pierre Boland, a scientist and one of three Canadian commissioners on the International Joint Commission, the U.S.-Canada panel that monitors border issues, including management of the Great Lakes. "Maybe there isn't now. But, as with many other things, you start taking a little here and a little there and eventually you find out that you're taking a lot of water. If you take enough, you end up lowering the level of the Great Lakes."

A point not often understood is that only about 1 percent of the water in the Great Lakes is replaced every year through the natural water cycle, such as feeder rivers and rainfall. The other 99 percent is fossil water, from the melting of glaciers about twelve thousand years ago. If you start to affect the flow-through, you start to mine the system, and the whole system starts to change.

But what does it take to affect the flow-through? The numbers are always tricky, and forecasts trickier still. In 1977 the Canadian Environmental Law Association issued an analysis of the Great Lakes system that forecast a drop of 25 percent in the flow-through in less than forty years. Climate change and "outrageous overuse" of water, mostly by agriculture, were blamed. I had already heard from Eugene Stakhiv, the Army Corps of Engineers researcher, who had noted that Canadian estimates of changes to the lakes were notoriously inflated. Of course, he was talking about climate change forecasts, not about overuse. But Canada's federal Department of the Environment greeted the CELA report with equanimity, pointing out that in 1977 the Great Lakes were "at record levels, excessively high." No need to worry; come back and see us in forty years. And, indeed, halfway through the forty years, nothing much had changed. The levels reached record lows in 1999, but no one thought it part of a trend.

The free trade argument over the Nova Group scheme was straightforward. We sell all our other natural resources, don't we? We sell our gas, our oil, our nickel, and our gold. We sell the renewable stuff, too — lots of trees go south to be sliced and diced and used to build houses. We sell our fish, or we used to when we had any. Why not water? Every barrel of oil and kilo of copper pulled from the ground and sold to foreigners is gone forever, yet these are respectable Canadian industries. Water resources, sensibly managed, are infinitely renewable and might make an important contribution to the national economy, yet export is tantamount to treason. And if an American company were in Nova's place? How do you use Canadian sovereignty to stop the Americans pumping away Great Lakes water? They could just pump it from "their" side. What are you going to do? Put up a net?

The anti–free traders, to whom NAFTA was in any case anathema, pointed out that under the agreement's Chapter 11, the nondiscrimination clause, once a permit has been issued — thus recognizing that water is a tradable commodity — it becomes impossible to deny equivalent permits to anyone else who wants one, and for whatever quantity. In fact, even banning the sale of water had the same effect. That, too, recognized that water was a commodity.

So what? the free traders asked. If you price it right, exports will be modest. And you'd have a good chance of making a decent buck.

But under the rules, the opponents answered, you can't discriminate against foreign firms. If you charged a "decent buck" for the water, communities across Canada, used to getting water free from the country's abundant resources, would have to start paying.

CELA believes that Canada caved in to American pressure when it failed to exempt bulk water sales under NAFTA. In a way, how-

ever, the story is even odder than that. No less a person than Pat
Carney, Canada's trade minister during the negotiations, seems to
have believed that water had been exempted and was puzzled to
find out it had not. The story was told in November 1993 by Brian
McAndrew, a *Toronto Star* reporter:

> It [the tip-off that water was not after all exempt] began with a ques-
> tion to a senior policy adviser to Pat Carney.
> "It's exempt, it's right there in black and white," the adviser said.
> But after trying to find the reference in the text, the adviser came
> up dry. "I don't know what happened. We discussed it. It should be
> there."
> The next tip-off came when Carney was tossed the same question
> during a constituency meeting three months later.
> "Water is exempt from the deal — it's right in the agreement," she
> replied. She too was asked to point out the wording in the text. After
> consulting an aide, she said, "it was there."

An editorial in Toronto's *Globe and Mail* found the flap reminis-
cent of General Jack D. Ripper in Stanley Kubrick's *Dr. Strangelove,*
who seemed an eerily Canadian figure in his obsession with foreign conspiracies "to sap and impurify our precious bodily fluids."

> Opponents have represented the Lake Superior proposal as the first
> tiny breach in the watertight dike, a breach that, once made, will be-
> come a rushing torrent, as thirsty Americans use trade instruments
> like NAFTA to plunder our hydrological resources, while water wor-
> shipping Canadians stand by, impotent. This image is wrong, on sev-
> eral counts. Canada already sells its water resources to foreigners,
> and not just in obvious but trivial forms such as beer, soft drinks and
> bottled water. Our hydroelectric exports, for example, involve mas-
> sive disruptions in our "water heritage," including river diversions,
> dams, [and] inter-basin exchanges of water. . . .
> As for Canadians' stewardship of their water resources, it is not
> impressive. We are second in the world only to the United States in
> our per-capita consumption of water, taking three times more each
> day than the average German, for example. A third of Canadians are
> not even served by waste-treatment plants. With respect to water,

Canadians and Americans suffer from the same disease: We say that
it is priceless, but act as if it were absurdly cheap. Most North Americans pay far less for their water than even just the cost of supplying
it, cleaning it up and returning it to the environment. Yet subsidizing
water use is economically and ecologically disastrous. In fact, heavy
subsidization of water in the US is the cause of any water "shortages"
that may exist there. In California, for example, agriculture consumes 80 percent of the state's water, often growing low-value and
water-intensive crops in the desert. If everyone paid the true value of
the water they used, they would use less, freeing more for those who
need it.[8]

The argument came to naught when Nova, chastened by the public uproar it had triggered, withdrew its application, on the condition that Ottawa work with the provinces to ban all water exports.
Quebec's separatist Parti Québécois government, for its turn, announced that it would declare all water under its territory property of the state, which would until recently have been a very
French solution, except that France began privatizing its own water a few years ago. Quebec reserved the right to sell its water wherever it wanted, a communiqué said, in another largely ignored
snub to the federal government.

Early in 1999 the federal government, its nationalist hackles now
up, promised legislation to ban water exports via "supertankers,
pipelines or man-made trenches." The environment minister,
Christine Stewart, was clearly on the side of virtue. "We consider
this an urgent issue," she said.

Nevertheless, the actual policy statement she issued was more
measured than her press releases. Her notion was to call for a
moratorium on water exports until a national accord could be
worked out. The governing idea of the policy was to protect Canadian watersheds, and in this context (an environmental context),
the whole notion of NAFTA was neatly sidestepped.

The International Joint Commission itself entered the debate
with an interim report in August 1999 calling for a six-month
moratorium on exporting water from the Great Lakes. It was re-

sponding to low water levels in the summer of 1999, declaring, essentially, that there was no surplus water to spare. The Canadian government responded by calling the report "too permissive."

Nowhere in any of this, however, did the notion ever emerge that water was not — or should not be — anyone's property. Water is not "ours" or "theirs" but the planet's. We use water, and it passes on, and then it comes back to us. But it is not, surely, something we should either hoard or prevent others from using.

And so it was left to a letter writer to the *Globe and Mail*, Shirley Conover, to set the ecological record straight and remind the politicians, the editorial writers, the corporations, and the citizens what water really is. "Water is not a renewable resource," she wrote in the summer of 1998. It only seems renewable because it keeps falling from the sky. But that is an ecologically primitive way of looking at things. It may be common sense, but, as so often, common sense can be uncommonly ignorant. "Renewable resources can reproduce themselves, that is, living things such as trees, cows and people. Water cannot reproduce itself. Water is recycled by means of the hydrological cycle: evaporation plus transpiration by plants, to cloud formation, to rain and snow, back to plants, rivers and ground water, to the oceans and cycling around again by means of evaporation, transpiration and precipitation. *The hydrological cycle is an ecosystem service, a life-support system for all living things, including humans*" (my emphasis). By removing water from one basin to the next, the basin being the hydrological cycle's recycling unit, we are tampering with this life-support system, with uncertain consequences. The Aral Sea, as we have seen, is a grotesque example, but there are many others.

Nova may have retreated, but others are still beavering away on the subject of water exports. The Canadian Environmental Law Association, which may have been overhasty on the Great Lakes, nevertheless has a finely tuned sense of what is happening in resource industries, and believes strongly that the newly privatized water companies are gathering for the kill. The British and French water companies, behemoths in world water management, are prowling the world looking for customers. Lyonnaise des Eaux became the largest water company in the world after merging with another French company, Compagnie de Suez, in 1997. It now distributes water to 68 million people in thirty countries, and has contracts with half a dozen Canadian municipalities, including Montreal. The company has been actively pushing the trendy notion of public-private partnerships, which the cynical interpret as a way for private companies to get access to all that wonderful tax money.

A Canadian company, Philip Services Corporation, which runs the water system in the Ontario city of Hamilton, successfully negotiated a contract whereby the politicians, and not the company, would be held liable for "any problems rising from the management of the facility," a neat way of having your cake, eating it too, and never having to pay for it. The American electricity and gas company Enron has taken over one of the British privatized water companies, Wessex Water, and is looking actively at the Canadian market. Other companies such as USFilter (which started its life as a treatment facility until it was taken over by the French firm Vivendi for $6.2 billion) have already signed contracts to "manage" local water companies in Canada, and in at least one case have filed for status as a municipality under Canadian law to bypass the tax code. Jim Southworth, president of another company called Consumer Utilities, has expressed his dismay that the Canadian tax code seems, inexplicably to him, to favor public projects. He had hoped, he told the industry newsletter *Global Water Report* in 1997, "for a more level playing field on the 7 percent Goods and Services Tax, which is levied on private but not public-sector services." Many other water companies have expressed their dismay at the "slow pace" of privatization in Canada. CH2M Waterworks, for example, an arm of Colorado's CH2M Hill, points to the more than two years it took to get regulatory approval to build a water treatment plant in Halifax, Nova Scotia.

Other Canadian private companies are getting into the water

business too. "Bottled water," one of them told the Canadian Broadcasting Corporation approvingly, "sells for a dollar a liter. Gasoline sells for half that." His company has been diligently mining an aquifer in Quebec, depleting it so much that local farmers are now obliged to treat the water they give their cattle. A proposal for another water bottling plant in the little Quebec community of Ways Mills set off a citizen uproar that sent the chastened city fathers back to the bylaw drafting table to keep "predators" out.

In Flesherton, in the central highlands of Ontario, parched farmers faced with falling water tables were stunned when the Ontario government blithely gave a company called Echo Springs the right to take, at no charge, 176 million liters a year of spring water. For its part, Echo Springs expressed bafflement at the uproar. "We're a clean, pollution free industry," Echo Springs president Mark Rundle told the *Globe and Mail* somewhat plaintively. "We bring jobs to the area. Where's the problem?"

Even if CELA is exaggerating — which is doubtful, given the declared intentions of these new international water companies to dominate water and sewage facilities worldwide — it's certainly true that the sight of all that water cascading into the sea, unused and useless except to the salmon and the tree huggers, causes a haze of irritation to spread over corporate executives in the water export business. John Carten, a British Columbia lawyer acting for Sun Belt Water, Inc., a Santa Barbara company that wants to export water from British Columbia to California (and has spent over seven years in court to put its case, so far unsuccessfully), believes Canada's policy is not just wrong but immoral. "A blanket prohibition on the export of water by tanker is a cruel and inhumane response to the real needs of people throughout the world," he told *Globe and Mail* reporter Heather Scofield in May 1998. "Whenever the subject of the export of water arises in Canada, we go a little goofy," he said. "I would suggest that a properly managed water export industry would be a real benefit to Canada and the thirsty peoples of the world."

On the other side of the continent, a Newfoundland construction company bought Gisborne Lake, near Gander, and sought permission to sell its water (via tanker) to Asia. Again, no customers were revealed, and the economics, on the face of it, were implausible, but the application stayed on the books for a few months until the Newfoundland government under Premier Brian Tobin put it out of its misery.

And then there's the Global Water Corporation, a small Canadian corporation with a world-sized ego, which has signed a deal with Sitka, Alaska, to ship American glacier water for bottling in China. Global, too, laments the lamentable Canadian ban on water exports: How shortsighted! All that easy money, just running away!

Carten thinks Nova's proposal makes no sense. It is British Columbia that has the water — maybe 400 million acre-feet of it, nearly thirty times the Colorado. All this water could have dollar signs attached to it. Sun Belt's CEO, Jack Lindsey, estimated that his company would have pulled in $1 billion in revenue over five years had the British Columbia government allowed Sun Belt to export a mere half-million acre-feet a year. Carten's indignation is obviously growing. In December 1998 Sun Belt launched another court assault, claiming $200 million from the British Columbia government for "violating NAFTA" and failing to do the right thing by righteously trashing the aforesaid "cruel and inhumane response." I couldn't help thinking of the case of Joseph Main, an Alberta felon. Main, who was convicted one more time of robbery in 1997, was diagnosed as having a rare psychiatric disorder called polydipsia, which causes him to drink possibly lethal amounts of water, up to 25 liters at a time. He would make the perfect customer for Sun Belt's new exports, a perfect metaphor for America's insatiable appetite.

Americans and Canadians have a long history of more or less amiable relations on water matters, partly because the politicians of

both countries have seen eye to eye on the value of exploiting water and controlling recalcitrant rivers, and partly because Canada's
sheer abundance — or perceived abundance — of water has made
it difficult to generate much political heat over water matters.

In 1909 the U.S. and British governments signed the Boundary
Waters Treaty, which was designed to minimize what frictions did
exist, and created, almost as an afterthought, the International
Joint Commission, which was given the right of approval or re-
fusal of any transboundary water issues. A shopping list of other
treaties followed, including the Lake of the Woods Convention,
the Skagit River Treaty, the Columbia River Treaty, and the St.
Lawrence Seaway Project.

In the many decades since its creation, the IJC has become a
model for managing cross-border resources: its careful balance
and perceived objectivity have meant that its decisions are seldom
challenged by either government. Its amiable American chief
commissioner, Tom Baldini, explained its success in admirably
nonbureaucratic language. "What can happen is that one country
controls the source and can pretty well screw the other guys," he
said, after delegations from the Aral Sea area and the Lake Titicaca
dispute in South America had trooped to the IJC offices for advice.
"What we have, and what we tell the others, is that there has to be a
willingness on the part of all participants to reach a fair and equitable agreement. And it's difficult — water is such a precious commodity."

The IJC has helped this willingness along by maintaining a rigorous objectivity, and by ensuring that the asymmetry in the U.S.-
Canadian power balance is not reflected on its boards. The chief
Canadian commissioner, Len Legault, maintains that objectivity is
built into the IJC's very existence. "We take a solemn affirmation
of objectivity," he says. "We are an independent body, and we
maintain strict impartiality. Commissioners represent the commission, not the governments who appoint them."

Late in 1998 the two national governments expanded the IJC's responsibilities. It would establish a chain of local transborder watershed boards with a much broader mandate to manage, to make decisions, than is currently granted to local water control and pollution boards. The first such board has already been set up on the St. Croix River, which straddles the Maine–New Brunswick border. Another dozen are planned, from the Atlantic coast to British Columbia rivers that flow into Alaska. A second millennium project for the IJC involves rethinking emergency response procedures for cross-border disasters, such as the Red River flooding of 1997.

For several years there were rumblings from Ottawa that the IJC should be handed responsibility for the hottest U.S.-Canada issue of the 1990s: the dispute over the Pacific salmon. But the issue was resolved, sort of, in June 1999, when the U.S. and Canadian governments announced a plan to set quotas along thousands of kilometers of the Pacific Coast, and to create a conservation fund to help reestablish salmon populations. The deal was no sooner initialed, however, than howls of protest went out from fishermen up and down the coast, and Canadian fishermen were accusing Ottawa of selling out, again, to the Americans, again.

In fact, the IJC might have to turn its attention instead to updating the Columbia River Treaty, a more or less successful international management treaty now showing serious signs of strain. In essence the treaty, signed in 1961, was designed for flood control and hydroelectric generation, and permitted the building of three new dams (the Duncan, the Keenleyside, and the Mica) on the Columbia on the Canadian side of the border, with certain American entitlements to the power generated in return for much of the capital used to build them. In addition, the Americans were allowed to build the Libby Dam on the Kootenai River in Montana, which backed up about 70 kilometers into Canada. A series of codicils compensated Canada for lands flooded by the Libby, outlined the water entitlements of both countries, and provided for a cross-border payment schedule for power used when generated in the other country. For the first thirty years this arrangement worked well enough, but at the halfway point in its sixty-year lifespan, disagreements were emerging. Under the treaty, Canada consented to sell its full entitlements to downstream power generation at a fixed price to the Americans. In 1995 this agreement expired, and the Bonneville Power Administration, which runs the power generation schemes on the American side of the border, broke off negotiations with British Columbia, maintaining that the province's new demands and price structures were "unrealistic" — a nice euphemism for "gouging." The IJC, given its record, will probably solve this dispute fairly easily.

More seriously, the original treaty was driven by the power industry and left no room for environmental issues, such as the West Coast salmon fishery. An article in the *American Review of Canadian Studies* in 1997 put it this way:

> Nowhere in the treaty are there formal provisions for the integration of fisheries and other environmental concerns with existing power and flood control priorities. Although adjustments can be made . . . to accommodate multiple-use concerns, such changes must be compensated for if they reduce power production on the other side of the border. Unfortunately this narrow approach to decision-making [has] created a climate in which management decisions, which ultimately are taken by hydroelectric and reservoir operation entities in both countries, are viewed as "nonpower vs. power" trade-offs. On both sides of the border, this faces the primary river managers with a most controversial issue: the inherent conflict between their obligations under domestic environmental legislation, and obligations under the Columbia River Treaty.

In effect, the classic rock and a hard place: the rock is the Endangered Species Act in the United States and the Fisheries Act in Canada; the hard place is the way the system has come to depend on managed water for electricity.[9]

The treaty's framers were innocent of environmental concerns. In 1961 rivers were by definition to be managed. As the power generation industry will tell you, good management is what made us all prosperous. But, as the CELA will tell you, it's that attitude that got us into this mess in the first place.

The dreamers have had other plans for Canadian water, some of them grand on a scale that would impress even the engineers who have covered the American West with its net of pipes, canals, ditches, siphons, dams, and pumps, and who are used to shifting a trillion liters across a landscape where it is naturally absent.

It all started with an ecological catastrophe in the making, the infamous Love Canal, more a dumpsite for chemicals than a true canal. In the 1970s hydrologists began examining the ooze that had entered the canal and found it a toxic brew of PCBs and other chemicals too numerous to catalogue. Some of this pollution had already entered the groundwater and was seeping steadily toward the gorge of the Niagara River. There were alarming eyewitness reports, and for a while television crews were stationed along the lip, watching for breakthrough. What a great story! You could actually see the seepage along the escarpment wall. It would take only a widening fissure to create a clean-water emergency of massive proportions.

The cities of Toronto and Montreal, Canada's two largest, drew their water from the Great Lakes and St. Lawrence system, downstream from Niagara. A plan was floated by engineering consultants which took on, albeit briefly, a virtual existence in the media and the minds of provincial politicians. The notion was to connect the pristine waters of Georgian Bay, a deep and protected adjunct to Lake Huron, to the communities of the St. Lawrence Valley. Georgian Bay had the advantage of being upstream of Niagara, as well as upstream from most pollution sources, except for a few struggling communities north of Superior, such as Thunder Bay. The plan's proponents thought it probable that they would be able to supply Toronto with potable water in its hour of need — and charge a stiff price for the privilege.

After a while, when the seepage stopped and the scientists began sucking the muck from the canal upstream, the TV crews lost interest. So did the potential investors: without an emergency, there was no profit. Toronto continued drawing its drinking water from Lake Ontario, and when the pollution levels reached a medically scary level, the problem was solved by pushing the intake pipes deeper and further. In 1998, impressed with the cold temperature of the water being drawn up, Toronto mooted a scheme to extend a pipe even deeper, to the really cold water, kept frigid since the ice age, and "mine" that water — not for the water itself but for its temperature. Why not use the cold water to air-condition all of downtown Toronto through heat exchangers and export the resulting warm water downstream to Montreal? Let them worry about temperature gradients.

The end of the Georgian Bay Canal wasn't the end of the Big Idea. In fact, it was only the start. The province of Quebec had already been building a massive series of dams and reservoirs to construct the $35 billion James Bay Hydroelectric Project, which exports huge amounts of power to the United States. The project was developed at ruinous cost to the indigenous culture of the Cree Indians, many of whose ancestral lands were submerged. Robert Bourassa, premier of Quebec on and off for several decades, began in 1985 to push for an even more massive project. The acronym his planners gave it was GRAND, for Great Replenishment and Northern Development Canal Concept. It would turn Ontario's Arctic Ocean port of Moosonee back into a freshwater terminus. A dike would be constructed across its northern side, the mouth of the U; the rivers pumping millions of acre-feet into the bay would gradually transform it into a freshwater reservoir. What would happen to all that new fresh water depended on who you were talking to. An aqueduct to the Great Lakes seemed the most popular version, with perhaps a canal from there to the Midwest or High Plains states.

All these schemes paled by comparison with the North American Water and Power Alliance (NAWAPA), an even more grandiose plan dreamed up by Californians to dam most of British

Columbia's dozens of rivers and turn them backward into the
Rocky Mountain trench, a new storage reservoir about 800 kilometers long. The province contains the third-, fourth-, seventh-,
and eighth-largest rivers in North America and could easily spare
at least ten times California's most extravagant desires. Some of
the water would irrigate all the High Plains; other bits would go
to the Colorado and the Great Lakes and even the Mississippi
system. This one never got very far, not merely because of its cost
— no one got around to working that out, though it would probably climb to somewhere around half a trillion dollars — but because the ecological consequences would be, almost literally, un-
graspable. It would do as much damage to the environment as all
the water diversions combined in America.

No one now thinks these projects, or any like them, will be built.
The day of the massive eco-engineering scheme is over. There is
much more understanding of the damage such projects do. In
1998, as a signal of changing sensibilities, the forest giant Mac-
Millan Bloedel agreed to end clear-cutting once and for all, the
given reason being "customer demand." And when the West Coast
salmon fishery collapsed because of overfishing, pollution, and
habitat destruction, there was more talk of reengineering projects back to the way they were than of building new diversions.
Even in "super-natural British Columbia," to use the province's
own marketing slogan, things were going awry. In and around the
town of Kelowna, a new parasite called cryptosporidium was rampant in municipal water supplies; as many as fifteen thousand people were infected. No one was entirely sure how it got there, but
the most likely culprit was the runoff of cattle waste from farms in
the Okanagan Valley, British Columbia's most valuable farmland.

There is a good deal of pressure to repair the damage some of
the grandiosity of the past has caused. The U.S. government is now
"studying" — a nice evasion — the breaching of the three lowest
dams on the Snake River, to the delight of the environmentalists
and the fury of the farm lobby, which in early 1999 was planning

what it was calling a "massive pro-dam rally." A farmers' meeting in Pasco, Washington, was addressed by antienvironmentalist gadfly James Buchal, who believes, in the teeth of the evidence, that the "salmon crisis" in the Columbia River Basin has been concocted by federal bureaucrats intent on "gaining control of the Columbia River."

The 215-meter Glen Canyon Dam on the Colorado, and the 290-kilometer Lake Powell above it, may also soon be history, if a campaign started by the Sierra Club gathers enough political clout. The campaign is also something of a mea culpa for the Sierra Club, whose director, David Brower, agreed to let the dam be built in the 1950s in exchange for the scrapping of two other dams he considered worse. Lake Powell drowned several thousand American Indian ruins and obliterated the rapids in Cataract Canyon, along with some of the most startling riverbank scenery in America. All that can be restored, the club's analysis says. And the notion that the dam provides enough water for Salt Lake City and Las Vegas is false: it loses almost as much to evaporation as it gives to those cities, and more water could be provided more cheaply through simple canals. Not everyone agrees, however, and among the dissidents are some eight thousand people living in the town of Page, Arizona, which exists only because of the dam. A Page resident was quoted in 1997 as saying, "They ought to line up those environmentalists and shoot 'em, every last one of 'em."[10]

Nor does everyone necessarily understand the Sierra Club's desire for pristine wilderness. A Scottsdale, Arizona, company, for example, has proposed something it calls, with a sublime disregard for topographic logic, Canyon Forest Village, a 2.8-square-kilometer development near the entrance to the Grand Canyon, which would include more than five thousand hotel rooms and 46,000 square meters in retail space, and which would draw its water from wells struck into the deep aquifers that feed some of the canyon's most beautiful creeks and falls. There is opposition, mostly from businesspeople in neighboring towns, and from environmentalists, who rightly fear the unleashing of further hordes on the canyon's fragile ecosystem. But in this case the project will likely get built.

As the new millennium begins, we find the Mexicans hunkering down, hoping for more water; the Canadians hunkering down, determined on no less water; and the Americans in the middle. The Americans retain their trademark combination of reckless use (egged on by a corporate system that will, if necessary, pave over the wilderness it wants to protect so that ever more people can see it), and creative energy (using that same corporate system to fine-tune demand and supply, and conserve by the logic of efficiency). Denver's Chamber of Commerce has already pointed out that while irrigated agriculture is an industry pulling in somewhere around $200 million a year, tourism pulled in twenty-five times that — more than $5 billion. For the first time, conservationists have an argument more compelling than using the Endangered Species Act to save the snail darter or another obscure warbler: money. In some places, at least, conservation is demonstrably good for business. It will help raise tax money. It will help politicians get elected. It will, everyone agrees, help the land look better. Most certainly, if conservationism can make everyone a dollar, conservationism has a future.

15

The Chinese Dilemma

*China is not running out of water, except
in places where water is needed most*

I F THE RICH are different because they have more money, China
is different because it has more people. The implications for the
world food supply are causing sleepless nights among the statisticians. The United States uses irrigation to produce only 17 percent of its food, and thus, if it ever runs out of irrigation water, its
own and the world's food supply will be challenged, but not catastrophically. China, in contrast, uses irrigation to produce 70 percent of its food, and if it ever runs out of water, the results could be
disastrous. China would have to import food — more food than
the world grain market currently has in surplus. If the Chinese
bought massive amounts from a finite supply, world commodity
prices would go through the roof. China might be able to afford it,
but what about the poor countries in the world who feed themselves with difficulty now, when grain prices are low? Remember
the numbers: there were 1.2 billion Chinese by century's end, an
increase of nearly 700 million since the end of the Second World
War. Despite rigorous attempts at curbing fertility, most of them
unexpectedly successful, China will in the next forty years add to
its population the equivalent of the entire North American continent. If the Chinese ate as much fish per capita as the Japanese,
that would take the entire annual world fish harvest of 101 million
tons. If the Chinese consumed as many eggs as Americans, they

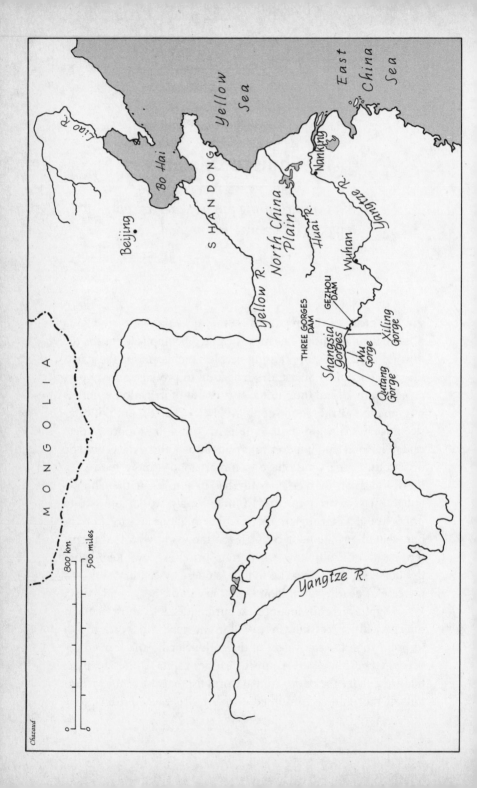

would require a flock of 1.3 billion hens, and reaching that goal would require an additional 24 million tons of grain, equal to Canada's entire grain exports.[1] The numbers don't just speak for themselves. They represent the most eloquent argument available for controlling human fertility.

The peasant farmers of Shandong Province on their new private plots were the first to go. In the early 1990s the level of water in their wells began to drop. They weren't aware of it yet, for Beijing had said nothing, but the water tables were dropping all over northern and northeastern China. The peasant farmers knew only that there was less water than there had been. In 1994 some of the wells went dry, and the following year hundreds more. All over the region, farmers and their families took time from cultivation to dig. They pushed the wells down another meter, two more meters, and found water. In 1996 their wells dried up again. The water tables on the North China Plain had dropped more than 4 meters in three years, and no one knew where it would end. For several years rainfall had been meager and rivers lower than usual, at a time when water demands were soaring. Extractions from river systems had historically been for agriculture, but China's astonishing economic boom, and its expanding industries and cities, were placing massive demands on finite water systems. According to China's own figures, between 1983 and 1990 the number of cities short of water tripled to three hundred, almost half the cities in the country; those whose problem was described as "serious" rose from forty to one hundred. By the year 2000, the Chinese said in 1998, Beijing municipality would suffer a daily water shortfall of 500,000 cubic meters, but it reached that dismal target before 1999 had arrived.

Recognition that a crisis is at hand is fairly recent. It was not until the 1980s that the region's water supply first began to seem fragile, and it is now only a few years since hydrologists and ecologists turned their attention to the problem. At Harvard, Henry Kendall, the late Nobel Prize–winning physicist and eminent scholar, lent his formidable intelligence to the question of world food supplies, which he believed was the critical issue for the next few decades. He steered me to Lester Brown, whose controversial "wake-up call" titled *Who Will Feed China?* had tracked the statistics and forecast that severe water shortages would force China to import food. His projections were later backed up by a study sponsored by the U.S. National Intelligence Council and performed by Harvard's Department of Earth and Planetary Sciences and Sandia Laboratory, a . . . well, secretive isn't exactly the right word, since it has its own searchable Web site. Perhaps "well connected" is the appropriate term. Sandia is famous for many things, among which are its simulation games. The Industrial Ecology Prosperity Game is a uniquely hard-nosed look at the effects of population increases, climate change, water shortages, and other ecological factors on the world economy.

Brown wrote with researcher Brian Halweil in the magazine *WorldWatch* in mid-1998:

A quarter century ago, with more and more of its water being pumped out for the country's multiplying needs, the Yellow River began to falter. In 1972, the water level fell so low that for the first time in China's long history it dried up before reaching the sea. It failed on 15 days that year, and intermittently over the next decade or so. Since 1985, it has run dry each year, with the dry period becoming progressively longer. In 1996, it was dry for 133 days. In 1997, a year exacerbated by drought, it failed to reach the sea for 226 days. For long stretches, it did not even reach Shandong Province, the last province it flows through en route to the sea. Shandong, the source of one-fifth of China's corn and one-seventh of its wheat, depends on the Yellow River for half of its irrigation water.

Although it is perhaps the most visible manifestation of water scarcity in China, the drying up of the Yellow River is only one of many such signs. The Huai, a smaller river situated between the Yellow and Yangt'ze, was also drained dry in 1997, and failed to reach the sea for 90 days. Satellite photographs show hundreds of lakes disappearing and local streams going dry in recent years, as water tables fall and springs cease to flow. As water tables have fallen, millions of Chinese farmers are finding their wells pumped dry.

If there were no more water in the rivers, what then? The theoretical answer is simple enough: some of the "users" would have to quit using it. The economics are even simpler. The people who stopped using it would have to be farmers. With a million or so liters of water, a farm would produce grain worth, say, $200. With the same amount of water, industry could produce goods that could be sold on the open market for twenty times that — and employ a good many more people.

Why is China running out of water? The answer is the same as for the rest of the world: it isn't running out. It's only running out in places where it's needed most. It's an allocation, supply, and management problem.

In the summer of 1998, one of the wettest on record, it was certainly not running out in the humid south. The Yangtze, swollen by sustained downpours, reached its highest level since 1954 and killed more than two thousand people in its rampages across the lower valley. That more people didn't die was a miracle. Most of the world's major watersheds are inhabited by only thirty or so people per square kilometer, whereas the Yangtze supports more than 220.[2] The industrial city of Wuhan escaped unscathed, though floodwaters had destroyed dikes upstream a day earlier, and it looked for a while as though more than 7 million people would be overwhelmed by the water. Chinese sources agree that the flooding was exacerbated by deforestation efforts in the upper basin, and by dikes that prevented the water reaching its natural flood plain, ancient wetlands now drained for agriculture. The periodic Yangtze floods were one of the given reasons why the Chinese government is spending more than $24 billion and resettling more than 1 million people to build the Three Gorges Dam. The other given reason is to supply water to the thirsty north.

The Chinese have a dilemma: three-quarters of the water is in

the south, while three-quarters of the farming is in the north and
northeast. The south includes the Yangtze, and about 700 million
people. The much drier north, inhabited by some 500 million people, includes the Yellow, Liao, Hai, and Huai rivers. The great west,
like America's, is largely desert. But there is no equivalent to California or the Pacific Northwest, no Washington and no Oregon,
only aridity. And what rain does fall in China is highly variable,
with little stability in supply, exaggerating the flood-drought cycle.
Chinese farmers depend on the spring rain, but it is particularly
erratic. Along the Yangtze variability from year to year is around 45
percent, and in North China it is more than 50 percent. East of
Beijing the variability index can exceed 80 percent.

The Chinese are not afraid of heroic engineering works, as the
Great Wall and the Three Gorges Dam attest. If further proof
were needed, look at the National Environment Protection Bureau's solution to the appalling pollution problem in the coal mining town of Lanzhou in central China. To clean up the pollution,
the bureau never contemplated cleaning up industry first or bringing in controls of any sort. No, the preferred solution was to ex-
pose the town to the prevailing winds by dismantling the 300-
meter Green Mountain to the east — bringing new meaning to the
phrase "letting in a breath of fresh air." Still, the absence of the raw
material — natural water — means that there is no prospect that
China could ever emulate the engineering works of the American
West or seek to make the deserts bloom, at least with its own water.
All the water is needed elsewhere.

But what about bringing it in from elsewhere — an "elsewhere"
that in China's case could only be Siberia?

In the late 1960s, when the Soviet Union and China almost came
to a shooting war along their long frontier, one of the issues was
water. The Chinese (at least this was the Russian view) had been
staring at their maps too long, and had been hypnotized by the
arid browns of the Gobi and the endless green of the Siberian tundra. It would take heroic engineering works to divert any of the Siberian rivers southward to make the desert bloom, but the Chinese

were famous for their long-term thinking and their indefatigable energy. Why would remaking a large part of their continental landmass deter them? Indeed, the tightly controlled Chinese media fanned these anxious flames by speculating more or less openly about the Soviet Union's eastern reaches, referring to them in code words easily understood in the Kremlin, where a similar language of obfuscation was readily practiced. "Our northern resource" was a phrase that came into currency in the early 1970s; this, together with the sudden appearance of Chinese maps with the border mysteriously omitted, was more than enough for the Russians, who began issuing increasingly bellicose threats. But it was all for nothing. The notion made no real sense apart from great-power rivalry. Why would China risk war for tundra of uncertain value when it had plenty of water of its own, even though it was in demographically vexatious places? And so it has proved. Beijing's eyes are turned southward now, not to the north, and the Siberian rivers continue to flow into the frozen Arctic, as they have done since the last ice age ended.

China's supply problem, as we saw in the chapter on pollution, has been gravely compromised by the quality of the water that is still available. When almost 80 percent of the rivers contain water unfit for human consumption, there is a serious danger that irrigated crops in the worst areas might poison the population instead of nurturing it.

China has not always been forthcoming about this or any water matter, any more than its government is transparent on anything it does. But there are strong signals that a crisis is predicted. In 1988 a new Water Law was passed, and the agencies responsible for water, soil conservation, and other regulations were consolidated in a bureaucratically superior Ministry of Water Resources, which reports, as other senior ministries do, directly to the State Council. The ministry controls water policy formulation; strategic planning, including flood control, water pollution, and wastewater controls; economic regulation; and conflict arbitration. Sensibly, the subministries were organized by river basin, each responsible for a major river system. Under them were the conservancy commissions, which looked after tributaries. The new scheme was designed to deal with what the announcement delicately called the "conflicts and shortfalls" of the older system, a bureaucratic nightmare in which no one knew who was in charge. That nothing had been done about the dikes and flood plains of the Yangtze by 1998 was not a reassuring sign that the new bureaucracy was any more flexible than the old, and late in the year the subministry in charge of the river was reduced to issuing innuendoes and criticisms of the local political bosses. River basin management was all very well, but political satrapies hardly ever coincided neatly with basin boundaries, and it was difficult to get anything done.

For the Yangtze, southern China's major water resource, shortages were furthest from anyone's mind in 1998. On the contrary, questions of how to manage and where to put the massive amounts of floodwater were preoccupying everyone. The Three Gorges Dam, which the government devoutly hopes will curtail rather than exaggerate the flooding, as some environmental groups have suggested, won't be finished for several years, if ever, and its conduits to the north won't be ready for at least another decade. Even in bad years, no one is forecasting water shortages in the Yangtze Basin.

For the northern rivers, it's another story. All of them are stretched to the limit, and already the water supply for 200 million people is assured only through unsustainable mining of groundwater. The Yellow River, as we have seen, already dries up in bad years before it reaches the sea. Development projects scheduled for its upper reaches, including hydroelectric schemes and a canal to Mongolia, will make matters considerably worse.

Lester Brown's thesis about China's looming grain shortages, first published in a WorldWatch paper and later in a book as part of the WorldWatch Environmental Alert Series, was indignantly denied by the Chinese, but the very vehemence of their denials cast doubt on their accuracy. But if you read Brown's argument carefully, it is far from an anti-China diatribe. On the contrary, he acknowledges China's substantial achievements in water management and suggests that China's problems are a factor of its successes, not its failures. Rising standards of income are leading inevitably to rising demands for water. He points out that the same thing has happened in other countries where development has been rapid, such as Japan, and there is no reason to believe that China will be immune. At the same time, more and more agricultural land is being taken out of production, partly because of competing demands for water, but also because the expanding and ever richer cities demand more space — for factories, housing, parking lots, and roads. It is politically necessary for the Chinese to create jobs and provide sanitary water supplies for the millions of citizens who live in the economically backward interior provinces, but giving in to their demands inevitably means increasing shortfalls in the amount of water available for irrigation downstream — water used in producing China's food.

The economics of water supply are compelling. In California it is a given that industry makes more productive use of water than farmers do; the market value of goods produced by water use in industry is fifty times the market value of food produced with the same amount of water. California found that from diverting water from irrigation to industry, the economy benefited, more jobs were created, and the tax base soared. Of course, food production went down, but food could always be imported. In China the same dynamics are at work, only more so: China needs to provide jobs for 15 million new workers every year. Brown's figures show that the demand for water by industry will increase from 52 billion cubic meters to 269 billion cubic meters by 2030: "In other words, non-agricultural uses that are now straining the system by drawing only 15 per cent of the supply would multiply nearly five times, while the agricultural needs now taking 85 per cent would have increased as well. Obviously, that can't happen."

When farmers lose groundwater or are cut off from irrigation sources, he says, they either survive on rainfall or they go out of the farming business. And food production declines.

Brown's analysis, say those who are familiar with the world's food supply, is fine as far as it goes. But he has made an assumption that may not be borne out by events: he assumes that the revolution in productivity brought on by the Green Revolution has run its course.

But after the Green Revolution, the Gene Revolution. A source at the Consultative Group on International Agricultural Research (CGIAR), an international clearinghouse for food-related science, has suggested that substantial gains in productivity are still possible.[3] "The bio-technology companies are re-inventing nature," he told me. "They are tinkering with the basic structure of plants and animals, developing nutritious food grains with high yields that should be able to thrive in arid conditions.

"What happens, though, when farmers can grow more food with less water and with a tenth of the labor? What happens is this: you get, finally, the complete industrialization of farming. Millions of Third World peasants will be out their livelihoods, no longer necessary, their lives redundant. They will be forced into the cities as slum dwellers, making that problem worse. Is this really what we want? You can look neither at water nor at food in isolation of other systemic problems."

PART IV

What Is to Be Done?

16

Solutions and Manifestos

If you're short of water, the choices are stark:
conservation, technological invention,
or the politics of violence

THEY'RE SORT OF like condoms," the guy from the International Water Resources Association was saying, "only bigger, a lot bigger. Huge. Like giant ribbed ticklers —"

The phone line wasn't so good. "Sorry, what was that?" I asked. "Ribbed what?"

"Ticklers. You know, like some women don't like the smooth condoms much, they prefer the, ah, well, ribbed ones because —"

"Yeah, okay," I said. I got the point. "But why are these bags ribbed?"

He sounded impatient. "Hydrodynamics. It's obvious. Why do fish have fins? If they weren't ribbed — if they weren't perfectly balanced, perfectly streamlined — they'd waste incredible amounts of energy twisting in the water."

"How big? How much bigger?"

He laughed. "A lot. Typically 650 meters long and 150 wide at their widest. You know how long 650 meters is? Nearly seven football fields. . . . These things are big. More than a football field wide. They've a draft of about 22 meters. They can carry a million and three-quarters cubic meters of fresh water each time. That's a lot of water, a million and three-quarter tons of the stuff. . . . But why don't you ask Jim Cran about them? He's the guy pushing the idea."

These "condoms" he was talking about were a modern spin on an older idea. Toward the end of the Second World War, smaller and more primitive versions, really just big rubber bags, were used for the high-risk activity of towing aviation fuel through the ocean to where it was needed. High risk, but perhaps still smaller risk than transporting the same amount of fuel in tankers during wartime: the tug needed to tow the bag left a much smaller sonar trace and was less likely to be thought worthy of attack. The bag itself was virtually invisible to sonar or, later, to radar, since its density wasn't much different from that of seawater itself. Of course, once detected, it would have been pretty easy to sabotage. One neat harpoon shot would have done it.

Later, a European consortium got the idea to use the technology to take fresh water where it was needed and tried an updated version. It wasn't "aquadynamic"; the closest marine model was a form of sea serpent. When it was towed through the water, it thrashed about, leaving a huge wake.

James Cran calls his "Medusa bags." His model was a jellyfish. Jellyfish don't thrash. They sail serenely through the water, generally unaffected by storms. But the Water Resources guy was a little premature, it turns out. Bags the size he outlined haven't been built yet. They exist, though, in Cran's head.

"And in any case," he said, when I called him at the Medusa Corporation office in Calgary, a long way from the sea, "they're not ribbed but strapped. Our original notion was simply a large polyester or nylon bag, a high-tensile woven polyester with plasticized PVC coating on both sides. We did a 5,000-ton prototype and deployed it in Howe Sound, off Vancouver, where we promptly pulled the front off it with our tug, proving that we still had something to learn. We did a redesign. We placed fabric straps every few feet to distribute the pull longitudinally."

"Oh!" I said, "the ticklers!"

He didn't seem to hear me. "Then we did a ten-meter latex model — talk about your large condoms! — and found something interesting. When it was towed through the water, it was totally unaffected by wave motions. The motions simply continued on right through the bag and its contents. It was, after all, just water, like the ocean, with almost the same density. This was wonderful: Medusa would be the world's first cargo-supported vessel."

The idea had come to Cran after a visit to friends in San Francisco, during one of California's periodic droughts. "After dinner I went up to use the bathroom. 'Don't flush!' they told me. There wasn't enough water. Not enough water in a city like San Francisco!"

Cran had been an engineer in Alberta's oil patch, but when the oil business fell on hard times, he turned his talents to designing software: "Spreadsheets and other products. In that way I became a project planning specialist. I had the software but not yet the projects. So I had to invent the technology to fit the projects that fitted my software. My first notion was a sort of hockey puck–shaped bag, small, about 400 cubic meters, being towed very slowly, and from that I progressed to our current notions, bags of a minimum capacity of 100,000 tons, but more likely a million tons, or a million and a half. The science proves out. The specific gravity difference between fresh and salt water is enough to keep us on the surface; there is a 2.5 percent gravity difference, which gives us a 2-foot head in pressurized bags, very streamlined, with a low rate of curvature." He gave the same dimensions I had heard earlier: 500 meters long, 150 wide, with a draft of 22 meters.

"We use ordinary tugs that proceed very slowly, at 2 knots — a speed, incidentally, that water engineers have found best for moving water through conduits, too. Our bags have a very large volume-to-area ratio, so the energy consumed per kilometer of movement is very low."

Cran took his idea to the Los Angeles Metropolitan Water District. "They pissed all over it," he said, unconsciously using a water metaphor. But more recently, the American West had once again been making anxious noises about running out of water. The droughts of the early 1990s notched up the rhetoric, and although in 1998 El Niño nearly washed the state away, there were still farsighted water companies that wanted to secure new supplies. Alaska's governor, Walter Hickel, floated the scheme of tapping the Copper River and pumping it to southern California in an undersea pipeline. No one laughed, although the estimated $100 billion price tag made it, almost literally, a pipe dream. The water would cost $3,000 per acre-foot, which translates into $2.43 per cubic meter. You can get desalinated water for that, and Cran could deliver for a fraction of the price.

Other water delivery companies with similar ideas were already talking to potential buyers in the Middle East. "It's not well known," one of them told me, "that there is a corner of Oman, near Salaleh in the south, that gets washed by monsoons. What a shame to let that water get away. Why not use bag technology to tow it to where it is needed?"

Dismissed in California, Cran went to Israel, through contacts with the Montreal Bronfmans, and got to Shimon Peres, a man who loved to put things together. "I think he saw a chance to put Israel and Turkey together, so he was interested. Why shouldn't Israel, which needs water, get it from Turkey, which has lots?" Everyone knows the Levant — Israel, Jordan, parts of Lebanon, Gaza, Sinai — needs water, and Turkey has about as much as Scotland, much of it rising in the mountains, and very clean. That's where the Manavgat River scheme came from.

In 1989 Cran arranged for the Israeli water commission to visit Turkey. "There was magnificent water there. On the Manavgat, two huge lakes in the mountains, with essentially no exit except one through a rocky channel. Lower down the town of Manavgat dumps its sewage, untreated, into the river, but until that point it is very clean. There is no farming upstream, and farming is a huge problem for fresh water." Tahal, the Israeli water agency, did a study of the Turkish situation and pronounced it sound. So was Cran's bag technology.

Israel asked Turkey for a letter of intent — some proof it was serious about sending water. The letter never came. Somehow, Colonel Qaddafi of Libya had heard about the project. There were all kinds of Turkish engineering firms working on the Great Man-Made River scheme, and they would not be paid, he insisted, not a penny more, unless there was a guarantee that no water would be sent to Israel or anywhere else without a comprehensive peace plan for the region. Turkey capitulated — almost.

Cran told the story: "Okay, they said, we won't send any water to Israel until then, but still no one can stop us building a terminal, a Manavgat Loading Terminal, for exporting water when the time comes." He snorted at this notion. This terminal, he said, was a joke. "I think, at last glance, it had cost around $120 million. DSI, the Turkish state water organization and the largest constructor of dams and irrigation works in the world, was asked to build it. But a terminal for what? No one told them. So they built the terminal for supertankers, despite the fact that no one had ever costed out water carried by supertanker, and there were no clients and no suppliers. Nor was anyone building a supertanker unloading terminal, in Israel or anywhere else. Worse, the Turkish terminal was designed to ship 180 million cubic meters a year (90 million treated and 90 million going for ballast). Israel, if it were to be a customer, wanted somewhere between 250 and 400 million cubic meters a year. So the terminal's potential was less than the smallest number the Israelis were looking at. The long-term demand in the area is around 800 to 1,000 million cubic meters. So, despite the cost, the terminal is just a toy."

Then came the famous handshake on the White House lawn, and Peres called on the Turkish foreign minister for that elusive letter of intent. Negotiations crept forward. Turgut Özal, the president of Turkey, visited Jerusalem and announced that his country could deliver water via the terminal there at $1 a cubic meter. "Everyone laughed. This was ridiculous. The Israelis had expected 25 cents, tops. Medusa can deliver the water for 20 to 25 cents a cubic meter — transport costs of around 10 to 15 cents, the rest for loading and unloading. And then Özal put forward his Peace Pipeline proposal. That would deliver water at $4 a cubic meter, even then the very top price for desal water. The whole thing was just a make-work scheme for Turkish engineers."

The idea is not dead but dormant. Israel, though tempted, said no — the Israelis don't really want anyone else to control their water. The 22-meter draft of the bags, not that different from the draft of the new supertankers, meant that the bags would have to be filled and emptied at moorings a few kilometers out to sea, served by submarine pipelines and pumping stations. There was some discussion about superpower delivery guarantees, but no one could persuade the Americans, the only superpower left, that this was a good idea — and in any case, American participation would be a challenge to every saboteur in a region where saboteurs are thick on the ground. Perhaps, it was suggested, Turkey could deliver the equivalent of two years' supply before payment was demanded, the water to be "stored" in Israel's aquifers. That would give everyone plenty of time to look for alternative sources if the Turks, for whatever reason, threatened to shut off the faucet. It was all becoming too complicated.

On an April morning in 1998, a Japan Airlines 747 lifted off the field at Osaka, heading for a primitive little airstrip that had never seen a jumbo before: Cambridge Bay in the Canadian Northwest Territories. The plane was empty, except for a few engineers and a number of strangely large compressors. The payload on the way back was to be chunks of Arctic ice. The plane had been hired by a Japanese sake maker, and the ice — pure Arctic ice, pristine and unsullied by the human propensity to pollute everything it touches — was to be used as a marketing device for a new brand of liquor called Arctic Pure. Never mind pricing of a dollar a cubic meter. This would be more like a dollar a liter.

A month later, on the other side of the continent, a small tour company was set to make money from "the last of the unsullied rivers," the pristine and beautiful Feather River in Labrador, one of the few remaining places on earth that has never been logged, mined, used by humans, or polluted by factory or farm. That there was anything incongruous about helicoptering people in and letting them tour the river in gasoline-driven boats to experience the fabulous cleanliness, the company's promotional literature didn't say. There were also plans afoot to bottle the water, to take it to the restaurant tables of the Western world, on the no doubt correct theory that people were increasingly doubtful about the purity of European bottled water. Both these schemes — the Arctic ice to Japan and the Feather River scheme — are neat metaphors for one solution to the water problem: if you haven't got enough, go fetch more.

Jim Cran's Medusa bags and Alaska's notion to pipe water to California are really just extensions of this approach. So are the plans occasionally mooted in St. John's, Newfoundland, and along the Alaskan coast to lasso icebergs and haul them south, chipping them up into tanker trucks before they melt (this was at least being thrifty: the water would melt into the sea in any case, so why not let it melt where it counts). So are the extravagant notions to return Mississippi Delta water to High Plains, Texas, or the now abandoned Russian projects to divert north-flowing Arctic rivers such as the Ob and the Lena southward, either through the Urals into the Volga Basin or toward the Aral Sea. So too were the ever more grandiose plans for "solving" North America's water problems: the GRAND Canal scheme to divert water from James Bay into the Great Lakes, or the NAWAPA scheme to transfer water from British Columbia practically everywhere, including the Great Lakes, the Mississippi, and California. So too the Three Gorges diversion scheme to take water to hard-pressed Beijing, the last of the grand projects that is actually being built — maybe.

It's not, after all, as though this approach didn't work. It works in California.

Every year the state moves 14 trillion gallons of water, in directions mostly south, capturing it behind 1,200 dams on every river and stream of any size, before fluming it hundreds of miles, lifting it over some mountain ranges and pumping it under others, fitting the sere landscape with a caul of pipes, ditches, and siphons that irrigates an agricultural empire far greater but not unlike the one that bloomed in the deserts of Babylon and ancient Egypt. And if wealth has tamed the water, then the water has made California wealthier still. If it were a nation, the state's $1 trillion economy ranks it seventh among national economies, and in addition to providing 55 percent of America's fruit, nuts and vegetables, California has managed to become the sixth-largest agricultural exporter in the world by intensively farming a region that receives less than 20 inches of rain a year.[1]

It's long been a cliché that, in California, "water runs uphill to money."

River diversion projects will still be built — in some cases because there simply is no choice. The Okavango, the example I gave at the start of this book, is a case in point: you have to get water somewhere. But the day of the massive diversion projects has effectively ended. There are few places left in America or western Europe for damming river water. The barely ticking-over Central Arizona Project was among the last of the diversion schemes contemplated by U.S. water agencies. That there are few places left anywhere that are still worth damming is one of the signs that the water crisis is real: demand hasn't stopped rising, and will double in the next few decades, but new supplies are increasingly hard to find.

There are people who maintain, still, that all will be well, that there will be no water catastrophe. It's a minority view, but sunny optimists can be found. The Spanish hydrologist M. R. Llamas is one who believes that many diagnoses of doom and forecasts of crisis have more to do with ignorance than real overuse of resources. He told the audience at a UNESCO conference in 1998 in Paris: "For example, we know water tables are dropping in many places. But we know nothing of the cycles. Maybe replenishment cycles are much longer than we assume, and therefore these overdrafts we're seeing now are just temporary. It is what I call a hydromyth. A hydromyth leads to the notion that ground waters are fragile, and shouldn't be developed. It's not always true that a falling groundwater table is a sign of something wrong. A steady state can take 100 years to arrive at.

"And take land subsidences. Extracting water does lead to ground collapses, it's true. But is this the problem it has been assumed to be? Mexico City has dropped 15 meters in places, which has damaged some buildings, but is it really the doom we have been told? In the Central Valley in California the land has subsided 10 meters in fifty years, without catastrophe — people have just adjusted. The real hydromyth is that all anthropogenic actions are bad."

He was received in stony silence. This didn't faze him at all. "Look," he continued, "the Saudis are using 20 cubic kilometers a year more than the replenishment rate. But why is it right to tell them not to use it now, just because they might run out in 200 years? Maybe we will have other solutions by then. And why should the present generation suffer for an unproved and assumed suffering ahead? Is Sudan's use of their own water a casus belli? Why shouldn't they use their water? Spain is the most arid country in Europe, so we will not plant unsustainably. But why should that give us the right to tell China what to do?"

When the stony silence continued, he said, with some exasperation: "I know of no place where overexploitation has caused catastrophe. Catastrophe is always in the future."

It's fair to say, though, that this is a minority view, shared by few hydrologists. The experts may not believe in some kind of water apocalypse — in some massive crashing of the hydrologically based human system — but they do believe the water crisis is real. Almost everyone believes, too, that there are really only three ways of coping with it: provide more water (either "make" it from seawater, or fetch it from elsewhere through engineering or diversions, as California does, as Medusa is trying to do, as Turkey's

Peace Pipeline would have done, as NAWAPA would have done in
spades); use less of it (through technological innovation, proper
pricing, good management, and conservation); or use the same
amount but for fewer people (that is, head off the crisis by sharply
limiting population growth).

Actually, there is a fourth: steal it from someone else.

Water survival strategy 1: *If you need more water, get more water.
That means you either import water from someplace where there is a
surplus, or you make more fresh water yourself.*

Is there really a demand for water towed around in humungous
plastic bags? Cran insists there is.

There are three main markets for delivered fresh water: the agricultural market, which generally can't afford water prices of more
than 10 cents a cubic meter; the municipal market, which can afford 50 cents a cubic meter or higher (in California the price is
generally around $350 per acre-foot, about 30 cents a ton); and the
homeowner market, the bottled water crowd, for whom the cost of
water is negligible — the container and delivery is what you're selling. Cran's market is the municipal one, and his main competition
is the desalination industry: "I figure," he told me, "we need a one-
third price advantage over desal to take account of the horrendous
political and bureaucratic problems of import-export. Our figures
show we can do this. Of course, we can generally only deliver to
coastal cities. But that's okay. So does desal. And there are real exchanges possible. If we deliver to Tel Aviv, it means Tel Aviv won't
be diverting Jordan River water from agriculture to the city, and
farming will have more water.

"Look at it this way. There are three places in the world where
there is massive movement of water through piping systems: in
California, in Israel through the National Water Carrier, and in
Libya's Great Man-Made River. All of these take water from the interior to the coast. This is ridiculous! Instead of California pumping water over mountain ranges and across deserts to take it to

L.A., why not leave it where it is and use Medusa water for the coast? One of our customers could be, for instance, the city of Las Vegas. They want Colorado River water now guaranteed to Los Angeles. Why don't they buy our water, 'give' it to Los Angeles, and simply keep the Colorado River?"

There have been other notions for carrying water through the ocean. One of those studied by the Intertanker group involved using superannuated oil tankers for transporting water. There are hundreds of single-hulled tankers rusting away in ports all over the world which are suddenly illegal for transporting oil; when the oil market collapsed in the late 1980s, the carriers tried to find markets for water but failed. Soon, also, dozens of newer supertankers will reach the mandatory retirement age. Why shouldn't they go into the water carrier business, for which safety standards are, for obvious reasons, not so strict? But the cost of the cleanup is estimated at $5 million per ship, and may in the end prove to be impossible. Benzene is not something you want in your drinking water.

And then there is the other tugboat solution: towing icebergs to where the water is needed. Greenland, Alaska, British Columbia, and Antarctica all have plenty of fresh water, and the icebergs are in any case breaking off and drifting out to sea. Why not capture them and take them to where they would be more useful? This concept has been widely explored. One of the proponents is a Saudi prince, Mohammed al-Faisal, who sponsored a study on the feasibility of towing Antarctic icebergs to Mecca. In the end, his group towed a block of ice around San Francisco Bay for a while and proved to their own satisfaction that no iceberg would ever survive to cross the equator.

Getting fresh water from the sea — essentially making new fresh water through desalinating seawater — is everyone's best technological hope for solving the looming crisis. After all, 97 percent of the water on earth is seawater, and 96.5 percent of seawater is really

fresh water in disguise, the other 3.5 percent the dissolved solids
that make it unusable for humans. Get rid of those solids, and you
can both drink it and irrigate your plants with it.

Amikam Nachmani, in a paper titled "Water Jitters in the Middle East," is typical of the political observers of the water problem
in calling desalination "the only realistic hope" for dealing with
freshwater shortages.

> The fact remains that Middle East water needs have increased con-
> siderably of late. The reconstruction of Kuwait, Lebanon, and Iraq,
> the influx of immigrants into Israel, and so on, all point to the need
> for a Marshall Plan for Middle East water. Is desalinization a practi-
> cal solution? Israel's official view, as presented in January 1992 to the
> Moscow multilateral water summit of the Arab-Israeli peace confer-
> ence, sees desalinization as the only long-term remedy for water
> poor areas like the Middle East.
>
> A desalinization project for 10,000 people costs the equivalent of
> one military tank; for 100,000 people, the price is roughly that of a
> jet fighter. Investing in desalinization of brackish water or sea water,
> or recycling sewage [for agriculture], is cheaper than attempts to
> settle disputes over available water sources, most of them already
> overused.

Nachmani quoted the political scientist Frank Fisher, who said
that "100 million cubic meters of water, which is the bone of
contention between Israel and its neighbors, is not worth a war: a
day of war costs $100 million, whereas desalinization of 100 mil-
lion cubic meters [of water] also costs $100 million." Similarly,
Nachmani maintained, "the diversion of water from one place to
another is much more expensive than the development of new
technology for cheaper desalinization. In addition, desalinization
plants create a sense of security by virtue of the state's ownership
of the resources and installations it considers vital (although Ku-
waiti and Saudi desalinization plants were targets of Iraqi attacks
during the Gulf War)."[2]

There are already more than 7,500 desalination plants in operation worldwide, some of them tiny. Two-thirds of them are in the

Middle East, 26 percent in Saudi Arabia alone. The world's largest plant, in Saudi Arabia, produces 485 million liters a day of desalted water; the Saudis and other oil-rich countries are using desalinated water for most of their domestic consumption. Israel's water controller, Dan Zaslavsky, estimated that it would cost about $2.5 billion to desalinate enough water to remove the water irritant from regional politics, a much higher figure than the $100 million of Frank Fisher but still considerably less than a war, even a minor war, when a sophisticated airplane can cost not much less than that.

Only some 12 percent of the world's desalinated water is produced in the Americas, and until the Yuma plant was conjured into being with federal subsidies to deliver clean water to Mexico, most of the plants were located in the Caribbean and Florida. The Californians have historically rejected desal; they got their water so cheaply (again through federal subsidies) that it hardly seemed worthwhile. As drought and groundwater overpumping heighten concern over water availability, however, desal plants are springing up all over the state.

In August 1999, Tampa Bay Water in Florida announced that it had hired a consortium to build a new reverse-osmosis desal plant that would produce 25 million gallons of fresh water a day, at a declared cost of 45 cents a ton. At second glance, the economics didn't seem quite as rosy. Over the life of the plant, the water would cost better than $2 a ton, but this is still cheaper than most other sources. The economics were possible because Tampa Bay is less naturally saline than the Gulf of Mexico, and because the plant would be built next door to a major power plant that would supply it with cheap electricity. Nevertheless, the price was such that it attracted a delegation from Singapore, which was planning a plant to produce 36 million gallons a day — at an average cost of nearly $8 a ton.[3]

The most beguiling — and massive in scope — desalination solution is the Red-Dead Canal, originally proposed by the American

Walter Clay Loudermilk and latched onto by the Jordanians. Jordan's engineers maintain that the canal could provide enough fresh water through desalination to solve Jordan's problems and enable Israel to pay what the Jordanians call its "water debts" to the Palestinians, Jordan, and Syria combined. Their analysis has been received skeptically by the Israelis but has never been rejected outright. Israel has floated an alternative, the Dead-Med Canal, which would, at least from the Israeli perspective, have several advantages: it would be entirely within Israeli control, and it would be shorter and therefore cheaper.

The Red-Dead plan would, in effect, extend the Jordan River south to the Gulf of Aqaba — except, of course, that the water would flow the other way, northward, for the 275 kilometers from the gulf into the Dead Sea. The water would be pumped from the gulf and along the Israeli-Jordanian boundary through the Arava Valley. It would have to flow uphill about 200 meters somewhere around Mount Edom, but from there it would begin a 660-meter descent to the Dead Sea, the lowest point on earth.

From here the planners wax ever more lyrical and enthusiastic, until their reports begin to sound like a feasibility study for some kind of Disneyfied theme park called Waterland. About 990 million cubic meters a year would hurtle down the valley in a series of great zigzags, creating boating lakes, swimming pools, and hydroelectric power as they went — enough power to drive the world's biggest desalination plant. About 40 percent of the descending gulf water would be rendered into potable water, and the leftover brine would decant into the Dead Sea, lifting its levels back to 1960s values and upping its tourist potential. The Dead Sea is much more saline than the ocean, and so too would be the brine left over from desalination. The tourists would float just as effortlessly as before.

Sure, it would cost billions, but once built it would pay for itself. The steep drop would generate sufficient energy to desalinate enough water to make the project self-sustaining. Well, that's the theory. Who pays for getting the water up the initial 200 meters is seldom discussed. The economic models are ambiguous, and the ecological consequences would no doubt be severe. There doesn't seem to be much political enthusiasm either.

Still, the proposal generates more interest than some of the alternatives. These include piping water from the Nile, siphoned under Suez, a notion that Egypt, already wrestling with water deficits, contemplated for political reasons (watering Sinai would be a potent political act) but cannot realistically afford. Turkey has proposed its Peace Pipeline to Saudi Arabia, and no doubt a branch line could reach out to the West Bank. And so on and so on.

In the end, though, it always seems to come back to desal — with or without a Red-Dead Canal.

The process of desalination, the removal of dissolved minerals (including but not limited to salt) from water, is not excessively complicated. Nor does its source water necessarily have to be the sea, though it usually is, desal plants taking ocean water through offshore intakes and pipelines or from wells on the beach. But desal plants can also use brackish groundwater or reclaimed (recycled) municipal water. Brackish water is not usually as salty as seawater, and is therefore cheaper to desalt, although for safety's sake wastewater desal plants usually also incorporate antipollution devices and biocides (typically chlorine).

The water treatment industry is fiercely competitive, and a number of different techniques have been developed, including reverse osmosis (RO), distillation, electrodialysis, and vacuum freezing. Only RO and distillation have so far been thought commercially feasible, but there is furious competition among research companies for new and different techniques. The commercial payoff for a reliable and cheap desal method would be incalculable.

The basic principle of RO is to pump the salty water at high pressure through permeable membranes or filters. There are other stages, such as pretreatment to separate out larger particles that would clog the membranes. And sometimes the water is passed through several stages of membranes before being declared "clean." The quality of the water produced depends on the pressure, on how salty the water is, and on the efficiency of the membranes. Since the membranes themselves are the heart of the process, what they are made of is usually kept a closely guarded commercial secret.

RO proponents like to point out that their plants take less energy to operate than distillation plants, have fewer problems with corrosion, and have higher recovery rates — about 45 percent — for seawater. Also, RO can remove unwanted contaminants such as pesticides and bacteria, and RO plants take up less space than distillation plants for the same amount of water production.

Distillation also has advantages. The plants don't have to shut down for cleaning or replacement of equipment as often as RO plants do. There's no need to pretreat the water because there are no filters to clog, and distillation plants don't generate as much waste.

Distillation, familiar enough from school science labs (or to anyone who has attempted a little homemade moonshine), is also simple enough: the water is heated, evaporated to separate out the dissolved minerals, and cooled back to water. As with RO, there are several competing technologies. The most common methods of distillation include multistage flash distillation (MSF), multiple effect distillation (MED), and vapor compression (VC). In MSF the water is heated and the pressure is lowered so the water "flashes" into steam. In MED the water passes through a number of evaporators in series, and vapor from one series is subsequently used to evaporate water in the next. The VC process involves evaporating the feed water, compressing the vapor, then using the heated compressed vapor as a heat source to evaporate additional water. Some distillation plants are a hybrid of more than one desalination technology.[4]

What's left over from any desal process is about one-third fresh water (up to 50 percent in very efficient plants) and two-thirds highly saline brine. The fresh water, though, is very clean, ranging from 1 to 50 parts per million of dissolved salts from distillation, and 10 to 500 parts per million for RO plants (drinking water standards are usually around 500 milligrams per liter, equivalent to 500 ppm). Because it's usually purer than most drinking water, it can be mixed with less pure water before distribution.

In May 1992 a desal plant owned by the city of Santa Barbara was analyzed by consultants for the California Water Commission. It produced 6.7 million gallons per day (MGD) of clean water, and generated 8.2 million MGD of waste brine with a salinity approximately 1.8 times that of seawater. An additional 1.7 MGD of brine was generated from filter backwash. This was figured at about 1.7 to 5.1 cubic yards per day of solids, equivalent to one to two truckloads per week.[5]

Energy is the big enemy of desalination, and a major barrier to its widespread implementation. Do we really want to purify water by adding greenhouse gases to the atmosphere? Distillation is a notorious energy hog, but even RO uses vast amounts of energy. "Desal is a tricky world," Jim Cran says. "There is some slippery economics out there. Some of the U.K. water companies, privatized by Thatcher and hungry for world markets, have engineering divisions offering desalinated water, and over long whisky lunches they'll talk about $1 a cubic meter for cleaned-up seawater. More honest people will admit that when energy and other costs are all taken account of, the real cost is somewhere around $2 to $2.20. For the Saudis, the world's major desalinators, the cost is closer to $4."

For Santa Barbara's desal plant to produce 7,500 acre-feet (9.25 million cubic meters) of fresh water would take 50 million kilowatts. To bring the same amount of water from the Colorado River to the Los Angeles Metropolitan Water District, fluming it over mountain ranges and along hundreds of miles of aqueducts, takes only 15 to 26 million kilowatts.

Price estimates for water produced by desalination plants in California range from $1,000 to $4,000 per acre-foot (81 cents to $3.24 per cubic meter). By contrast, in 1991 the Metropolitan Water District paid $27 per acre-foot for water from the Colorado River and $195 for water from the California Water Project. Non-interruptible untreated water for domestic uses in San Diego is purchased from the MWD for $269 per acre-foot; treated water costs an additional $53. The least expensive new supplies, other than desalination, would cost $600 to $700 per acre-foot. In Santa Barbara the development of new groundwater wells in the mountains would cost approximately $600 to $700 per acre-foot. Enlarging the Cachuma Reservoir, if it even proved feasible, would bring in new water at about $950 per acre-foot. During the drought of the early 1990s, the city purchased water from the State Water Project on a temporary, emergency basis at a cost of $2,300 per acre-foot.

Desalination techniques are getting better and therefore cheaper. Perhaps at some point they will become suitable for a mass market. The research, meanwhile, is ongoing. The Middle East Desalination Research Center, created in 1996 in Muscat in the gulf state of Oman, is where much of the more interesting work is being done. In April 1998 its director, Eric Jankel, put out tenders for projects that included the development of "small, home-use RO facilities" and a number of other innovative small-scale desalination techniques (along with, it's fair to say, a series of worthy but not so sexy projects such as "scale-prevention techniques for thermal desalination").

A British firm, Light Works, Ltd., was already involved in work for the center on what it called seawater greenhouses, essentially using the ocean as a heat sink to produce fresh water "by humidifying warm air using sea water, and then recovering pure water from the humid air using a condenser, also cooled with sea water." Running this process inside a humidity-controlled greenhouse had the added benefit of cooling and humidifying what one study called "the growing environment." The process could not only get fresh water from the sea but also grow plants in the very environment that produced the water in the first place. And a team from San Diego was working on solar-powered desal plants, using off-the-shelf technologies such as solar dish concentrators and a solar-powered variable electric motor drive. They hoped to be able to produce small portable systems for widespread distribution in Middle Eastern and North African countries, offering a "near-term solution to water shortages in remote locations."

Water survival strategy 2: *If you can't get more water, use less of it. Reduce demand. This can be done in three ways: by conservation; by pricing mechanisms; or by making existing water consumption more efficient through a combination of a new water ethic and skillful use of imaginative technologies.*

Eugene Stakhiv, the Army Corps of Engineers scientist, is no more a believer in a water apocalypse than the Spaniard Llamas. He believes passionately that the doomsayers are devaluing human ingenuity. Just look at the record, he says, how often prophets of doom have been proved wrong by some new invention: the hydraulic pump, or more productive seed varieties, more advanced farming techniques, or a revolution in technology.

These inventions can come from unlikely sources. Take the chemical giant Monsanto, for instance, once notorious for providing the U.S. Army with Agent Orange and now reborn, at least in its public relations, as a biotech company with a stated commitment to sustainable development. Hendrik Verfaillie, head of the company's agricultural division, told a Toronto conference in 1998 that solutions to the world's food problems lie in genetically engineered crops that produce more food per hectare with less labor, less water, and less fertilizer. His company estimated that world food demand would increase by a factor of three in the next fifty years. This would require another 38 million square kilometers of croplands. "If that were the case, we'd have to burn down the rain forest," he said, with considerable hyperbole. "We would have to eliminate all the wetlands and tax the environment in a way that would be totally unacceptable. Another possibility is to increase productivity by a factor of three. And one way to do that is through biotechnology."[6] Still, it never does to be overly credulous about corporate spokespeople and their public statements. There are other facets to biotechnology's newly kindly eye. Biotech companies have been diligently patenting seeds, which means that farmers, many of them poor peasants, who want to grow even traditional varieties might henceforth have to pay a royalty. Many of the reengineered seeds have a self-destruct mechanism built in so they no longer replicate, forcing farmers back to the biotech warehouses for the next season's crop seeds.

When you start to look around, it's hard to disagree with Stakhiv's high opinion of human ingenuity. For example, rainmakers (more accurately described as "rain-boosters") are having some success, mostly because a South African scientist, Graeme Mather, had noticed an odd phenomenon: emissions from a paper mill seemed to be making it rain (or, rather, rain harder). Mather analyzed the emissions and found two hygroscopic salts, potassium chloride and sodium chloride. Hygroscopic substances have a strong chemical affinity for water, and these were apparently making water droplets bigger and heavier, precipitating them as rain. He seeded other clouds with the same particles, released from flares under the wings of small aircraft, with positive — and measurable — results. But there are plenty of other examples of ingenuity at work.[7] For example, water is now being "stored" in deep aquifers: winter river flows are being pumped into the chalk aquifers underlying London, a response that has reversed the downward trend of the water table; Arizona, not yet needing its full Colorado River allotment, is storing the surplus in deep aquifers there. This replenishment has twin advantages: it doesn't divert water from its natural basin systems; and it minimizes evaporation, making the storage more efficient than surface reservoirs, especially in very hot, arid regions. And in Senegal, a Canadian-Senegalese joint venture is restoring water to the country's so-called fossil valleys, irrigating and repopulating land along 3,000 kilometers of dried-up riverbeds, refilling the valleys with water from the Senegal River without at the same time endangering native fisheries, agriculture, or navigation. But for me, a small and largely forgotten example of technological inventiveness, and one of the most beguiling, is the work of the Russian Feodor Zibold in the early 1900s.

Zibold was a "natural scientist," as they were then called, a philosopher of nature, a man consumed with a childlike curiosity about how things really worked. There were people like him all over Europe a hundred years ago, men of independent means and independent minds, all imbued with the kind of sunny optimism that goes with exploring where none other had been before. Charles Darwin is the most famous exemplar of this agreeable species. Zibold came later, but he was operating on the fringes of Europe, in the last tottering days of czarist Russia, and could be forgiven his dilatoriness. And, unlike Darwin, he was often spectacularly wrong, which only adds to his charm.

One day late in 1906, when he was taking the waters in the Crimea, he came across a local legend that the ancient Greeks, who had built an important provincial capital at Theodosia (now Feodosia, Ukraine), had mastered the morning dew: they had become so proficient at collecting and dispensing it that they had supplied the whole city with its fresh water, using neither well nor brook.

Zibold was captivated by this notion and determined to recover the secret. Once he started to look around, the evidence was everywhere. Lying about on the ground were stony tumuli, undoubtedly remnants of ancient dew collectors, and surrounding them were clay pipes, which had clearly been used for conducting the water to storage cisterns. Filled with enthusiasm and energy, he bullied the local agricultural community into helping him build his own massive dew collector, a stone reservoir 20 meters across, in its center a pyramid of stones and pebbles 6 meters tall. Halfway through construction he ran out of money, and it wasn't until 1912 that the marvel was finished. And, indeed, it worked: a yield of 350 liters per day was reported, not exactly city-slaking news, and not much water per ton of rock, but something.

In subsequent years the yield dropped, for no apparent reason, and by 1917 the local farmers had more on their minds than dew. They were more intent on surviving the Bolsheviks' maniacal enthusiasm for collective farms, which in fact worked a lot less well than Zibold's dew farm, and the famous dew collector was abandoned. The remnants are still there, however, and curators of the shabby little museum in town will happily show them to you and explain how Zibold's "condensers" were in fact ancient Scythian tombs, and his clay pipes dated back no further than the Middle Ages, when they had been used to bring water into town from a nearby spring.

And the point of the story? Zibold's work has inspired dozens of emulators, among them the Grenoble researchers Daniel Beysens, Irina Milmouk, and Vadim Nikolayev, who have developed workable dew collectors for use in areas where actual rainfall is scarce. In a paper they presented in 1998 to the International Conference on Fog and Fog Collection, they showed that their small collectors can produce "rainfall in the order of 0.1 to 0.5 millimeters per day, and are still open to improvement, the ultimate goal being continuous formation of dew, day and night. . . . Optimizing the parameters [might] lead to a new generation of condensers able to provide clean water wherever ordinary means cannot be used."

The fog conference is itself ample evidence of humankind's intellectual fecundity. It was held, appropriately enough, in the damp climate of a Vancouver summer. More than 140 scientific papers were presented, ranging from the mind-bogglingly arcane ("Fog Chemical Climatology over the Po Valley Basin") through the poetic ("Fog from Space") to the intensely practical ("Fog Water Collection for Agricultural Uses in the Darjeeling-Himalaya, India"; "Design, Construction and Operation of a System of a Fog Water Collector"; "Evaluation of Fog-Harvesting Potential in Namibia"; "Fog Collection as a Water Source for Small Rural Communities in Chiapas, Mexico"; and the Grenoble paper, which was called "Dew Recovery: Old Dreams and Actual Results"). Since fog, as one delegate put it, "is simply a cloud on the ground," harvesting drinking water from fog was a major theme of the conference. The consensus? A collector can generate up to 10 liters of water per collector meter of mesh per day.

And more on fog. In the Namib Desert, one of the oldest on earth, there is a dazzling array of creatures with idiosyncratic adaptations to the extreme heat and dryness: a lizard that hops from foot to foot to diffuse heat absorption; a beetle that curls into a ball to roll down dunes to conserve energy; a spider that spins a small cone-shaped web to track and condense dew; and a fog collector beetle. This little beetle's fantail is ribbed, and the fog condenses on the tail, drips down the ribs, and into a waiting sac. Scientists in California and in Israel have spent years studying and mimicking this ingenious invention. An experimental station in the Negev uses fog-drip irrigation for small-scale agriculture, creating cones of moisture-trapping fibers around individual plants. It is one of the many curious devices emerging from Israel's inventive antidesertification program.

Israel, which is mostly arid (annual rainfall of less than 200 millimeters), has always been in the forefront of water conservation techniques. The Israelis have been famously efficient at deploying drip irrigation, low-pressure spray irrigation, wind-trap funnels for moisture control, cloud seeding, and the rest of the panoply of arid region remedies. The Jacob Blaustein Laboratory has invented a hothouse cultivation technique, closed-cycle hydroculture where evaporation is recycled, which wastes only thimblefuls of water. Drip irrigation, a name that is self-explanatory, is fairly capital intensive — farmers need several kilometers of hose and hundreds of emitters per hectare of land — but it is truly efficient for certain kinds of crops. These techniques have spread to arid areas everywhere. In California the makers of high-end wines have learned to control the delivery of nutrient-enriched water to the nearest teaspoonful, and have even managed to use controlled water stress, which forces the plant into desirable growth habits by modest amounts of water deprivation at the right time, to concentrate the juices. On the fringes of the Sinai and other deserts, Bedouin, formerly nomads, are now employing drip irrigation on oasis crops. Low-pressure spraying, which can be used for feed grain crops, consumes 30 percent less water than conventional high-pressure spraying, and 60 percent less than conventional furrow irrigation. Laser leveling of fields has reduced irrigation water consumption considerably, in parts of California's Central Valley by as much as one-third. In the Sacramento Valley rice farmers have done even better, with protective screens on river diversions and the development of less water-intensive crop varieties. (They point out, rather smugly, that rice fields also serve as seasonal wetlands for migrating waterfowl.)[8] It's not just farmers: American industry, which began diligently recycling its water as prices went up, has seen its consumption drop by almost 35 percent since 1950, according to a survey released by the United States Geological Survey in 1998. (Overall American water use dropped by 9 percent between 1980 and 1995, according to the same study.)

In August 1995 the Israelis, bureaucratically incarnate in the Interim Secretariat of the Convention for Combating Desertification (INCD), got together with their Jordanian and Palestinian counterparts and came up with a fifteen-year plan to conserve water and combat desertification, budgeted at around $400 million. Their approach, flexible and inventive, was to turn the ecological disadvantages of drylands to economic advantage. Among other things, they analyzed the development of closed irrigation systems; the integration of flood-dependent production systems that build on rather than try to suppress flood pulses; the introduction of water treatment and solar energy systems that exploit radiation and heat from the sun; the cultivation of desert crops; forestation and improved management of grazing lands; the development of aquaculture to exploit both sun and saline water for the cultivation of fish, seafood, and algae; and the promotion of ecotourism, based on annual bird migration and the rehabilitation of endangered desert mammals. There was even, according to the Israel Information Office, a breeding program for camels and a camel study course, which among other things would use camels for transportation, draft, and tourism ("Camel races are popular in many Arab societies . . .").

There are aquifers containing substantial amounts of water of varying salinity levels underneath much of the Middle East and North Africa, from Morocco in the west to Saudi Arabia in the east. In the Negev Desert in Israel, salinity levels range between 3,000 and 6,000 parts per million of total dissolved solids. Learning how to use saline water for irrigation is a challenge. Major crops that are today irrigated with saline water include cotton, wheat, corn, table tomatoes, and melons. A number of secondary crops are grown with saline water as well. Research currently focuses on Bermuda grass, potatoes, grapes, and olives. The institute would breed and select crops for salt tolerance.

On the outskirts of the central South African town of Bloemfontein when I was growing up some forty years ago, there was a "sewage farm," evaporation ponds for the town's wastewater combined with irrigation channels to nearby cornfields. As schoolchildren we were dutifully bused out to see how it worked, part of a civics course that took us also to the municipal waterworks. I can't say that it did us much good: we were too squeamish to look closely, and the boys spent most of their time during the visit threatening to push one another in. In later years the "farm" was abolished. The city spread out beyond it, and the population increase, combined with more modern methods of chemical treatment in underground silos, meant the end of the older system.

This change was fairly typical of what was happening elsewhere in the world. The irrigation of crops with wastewater, even treated wastewater, fell out of favor. There were concerns about the spread of disease from consuming products irrigated with raw water from sewage systems. The "sewage farms" came to be seen as unsightly and unsanitary.

It took Israeli ingenuity, pushed by the country's water shortages, to bring them back into favor. In the mid-1990s Israel was already reusing 70 percent of its sewer water, safely irrigating 20,000 hectares of land. By century's end, reclaimed wastewater was supplying more than 16 percent of Israel's total water needs. This is what Sandra Postel calls the "elegant solution" of the agro-sanitary approach: partial treatment in wastewater lagoons and reservoirs, followed by irrigation, to solve simultaneously problems of pollution, water scarcity, and crop production in dry lands.[9]

In the American West, wastewater reuse is coming back into favor. Tucson, Phoenix, and Los Angeles all recycle portions of their wastewater. St. Petersburg, Florida, has closed the cycle altogether, reusing all its wastewater and discharging nothing to the rivers or ocean. The wastewater not only irrigates city parks and residential lawns but also has drastically cut the need to buy fertilizer. In El Paso, city engineers are pumping treated wastewater into a deep aquifer, "mining" it several years later and several kilometers "downstream," and using it to water parks and public spaces. As Postel puts it, "A major push by development agencies, governments at all levels, and private engineers to combine low-cost sewage treatment with irrigation could go a long way towards solving the vexing triad of pollution, scarcity and health problems now plaguing so much of the world."[10]

These solutions are not without their hazards. In the United States, sewage sludge, even after treatment, has on occasion been found to contain pathogens and toxic chemicals, including PCBs, DDT, dioxins, salmonella bacteria, and even lead, mercury, polio and hepatitis viruses, parasitic worms, asbestos, and radioactive waste. Some industry spokespeople have voiced concerns that the "beneficial uses" of sludge (which means using it as fertilizer) could introduce viruses into the food chain. Nevertheless, faced with few options for disposing of urban wastes, large cities are selling their "beneficial" sludge at a loss. New York City, for example, has paid more than $100 million to "fertilize" a ranch in Sierra Blanca, Texas, with 400 tons a day of possibly toxic waste.[11]

Technology, or human inventiveness, is one answer to the challenge of conservation. But on its own, technology is not enough.

Much of the current water wastefulness is political in its origins. Pricing and public policy — economics, broadly defined — are critical to any solution.

"The key to water is market economics," Maurice Strong was saying. "Too-cheap water subsidizes inefficiency. And there are too many perverse subsidies: public funds are being used for anti-public purposes." Strong, the secretary general of the Rio Conference on the Environment in 1991, a board member of Ismail Serageldin's newly formed World Water Commission, and an eminence in dozens of earth-friendly environmental groups (including his own Earth Council), was talking about a study the council had commissioned called "Subsidizing Unsustainable Development."

I told him of Sandra Postel's notion that conservation was humankind's last oasis, that the easiest way to find water was to stop losing it. Inefficient water delivery systems are hugely wasteful of water. In the Philippines and Thailand, more than half the "available" water simply disappears through leaks and rotting infrastructure, not to mention theft. Even in Jordan and Yemen, the annual water supply could be increased by one-quarter if the disappearing water was made to reappear.

"Yes," he said, "we have to bring supply and consumption into balance. And the most efficient way of doing that is pricing. Most farmers still use the old irrigation furrow method of watering their crops, although it is the least efficient way of doing it. Surge irrigation techniques, laser leveling of fields, all those systems can reduce water consumption dramatically. They are expensive. There is no incentive to introduce them unless water becomes expensive too."

Even in Israel, I remembered, this was a fact of economic life. Israeli farmers still receive substantial water subsidies, which have encouraged profligate water consumption and the growing of export crops with a huge water appetite, such as cut flowers and tropical fruit. And some studies have shown irrigation to be an ambiguous benefit. In some regions, dryland farming (rainfall-dependent cropping) might have yielded equal tonnages; and it has been suggested by some Israeli agronomists that perhaps the "blooming of the Negev" should never have been attempted. A recent increase of about 12 percent in the price of water caused a 10 percent drop in consumption, and a consequent drop in agricultural production. This decline seems to have had zero effect on the Israeli GDP; the freed-up water was gobbled up by industry, which has a greater economic multiplier than farming.

In Australia's Murray-Darling Basin, planners shifted from a command-and-control approach to water (useful in times of abundance) to a market-oriented property rights system that allowed "owners" to trade their rights with users prepared to pay a negotiated price. To no one's surprise, the new approach brought about a more rational and equitable allocation of the resource.

"In the Ogallala area," I told Strong, "they've been using new low-energy sprinklers. Water use is down in High Plains, Texas, by 44 percent. Some of this decline is by taking land out of production altogether. But not all of it."

Strong nodded. He knew all that. After all, he owned part of a Colorado aquifer himself, and, though not an American citizen, was once asked to chair the Western Resource Council's water committee. He rummaged in his briefcase and pulled out a sheaf of papers, clippings, and notes. "It's the same as with energy," he said. "If gasoline was priced to take account of its real cost, including the cost of remedying pollution, it would lead immediately to reduced consumption. So with water. What, for instance, about a water depletion tax? Why should the taxpayers subsidize the unsustainable withdrawal of groundwater?" He rummaged some more. "Ah, here it is," he said. "'Subsidizing Unsustainable Development.' Done for the Earth Council."

I took the paper home. Water was Chapter 2 of the study. There were the usual graphs of water-scarce countries and freshwater withdrawals, the usual rundown on humankind's profligacy, without which no environmental paper would be complete. "It is time for a reality check," the authors say.

Aquifers are being drained, rivers are drying up, more than a billion people don't have access to safe water, and vast tracts of irrigated land are being lost to salinity; over all, water is being lost in flood proportions and used inefficiently and for low-value purposes. And what are governments doing? Subsidizing this ecological vandalism, natural resource waste and economic perversity by selling water well below actual supply cost, much less [than] market value. The message of a subsidy is clear: don't worry about conservation or higher efficiency or recycling.[12]

The study then runs through the litany of abuses I had already become familiar with through Marc Reisner, Sandra Postel, and Peter Gleick: U.S. subsidies to agricultural irrigators amount to nearly $500 per acre (sometimes on marginal land worth much less than that); the total annual subsidy is probably greater than $2 billion; three-quarters of the benefits of the Animas La Plata Dam in Colorado go to farmers, who are paying only 3 percent of the cost; 70 percent of the farmers' profits in California's Central Valley — which is supposed to be the richest farmland in the world — came directly through taxpayer subsidization. And more numbers: in Australia, water charges are largely nominal in the Murray-Darling river basin, the major agricultural region; the forgone budget revenue from illegal connections to water mains in developing countries is around $5 billion; budgetary savings from fixing leaking pipes would be around $4 billion a year; subsidies and potential savings in the drinking water sector alone would be around $44 billion; cost recovery from farmers in developing countries is no better than 10 to 20 percent; "farmers in perpetually parched Tunisia pay no more than a seventh of the cost of the irrigation water they receive. . . ." The figures were familiar, and familiarly depressing.

What would happen if subsidies were removed? Wouldn't farm prices go up? Isn't the reason for the subsidies to maintain cheaper food prices in the first place? Wouldn't many farmers go broke? Wouldn't the poor be penalized most? The poor can't pay high prices for anything, and if water were treated as an economic good, priced to regulate demand, wouldn't the poor have to go without?

The answers given by André de Moor and Peter Calamai in their Earth Council report tally with theoretical studies elsewhere. No government has yet had the guts to remove water subsidies altogether, but where it has been done regionally or locally, there has been considerable demand elasticity: a 10 percent price increase typically yields a drop in demand of 15 to 20 percent. Metropolitan Boston saw demand drop by 30 percent when it increased prices and mandated water-saving plumbing fixtures. In Canada, where taxpayers typically subsidize water delivery systems to the tune of about $3 billion a year, residential water use under a flat-rate system averages 450 liters per person per day, and a mere 270 in those communities with a full user-pay system. A study in Egypt had already shown that a forced drop in water use would affect GDP only marginally. A similar study in Morocco, where 92 percent of water is used by farmers, with only a 10 percent recovery, showed that dropping most subsidies would cut water use by one-third — and affect GDP by only 1 percent. And the researchers predicted that the drop could be transformed into a surplus by liberalizing trade and by factoring in the ecological gains made by using less water. In reducing soil salination and waterlogging, more land would be kept in production for longer.

In fact, de Moor and Calamai say, the notion that the urban poor would be damaged most by an end to subsidies turns out to be incorrect. Subsidies mostly benefit the rich, since they are most likely to be connected to a public supply in the first place. Outside the developed world, the urban poor often have to rely on small private water sellers, and typically pay between $2 and $3 per cubic meter, twelve times the price of public city water. The same is true even of farmers. It is the wealthiest farmers who tend to occupy the best land and have access to wells and pumps. The most extravagant example of welfare for the wealthy is in the Central Valley in California, where millionaire owners of agribusinesses are subsidized by taxpayers all across the nation (while at the same time they rail against "big government" and "government interference").

The answer, the report suggests, is demand management instead of the supply-oriented system now in use. Charge not according to what people can pay but according to what the water is worth, taking into full account the development costs of delivery systems. Water has an economic value in all its competing uses, and sound water pricing will achieve more sustainable patterns of water use and generate the new resources necessary to expand services. For farmers, too, prices should be adjusted to encourage sustainable practices. If we stopped subsidizing water to plant water-thirsty crops alien to deserts or to irrigate pasture to raise cattle for beef (at a 50,000 to 1 ratio — 1 kilo of beef from 50,000 kilos of water), if we stopped doing the wasteful things for which the American West is famous, we would have water to spare.

Technology has its role to play here, too. Demand can be reduced through improved technology, in the cities through devices such as low-flow faucets and toilets, and in agriculture through better surveillance and control devices.

The newly privatized water companies in western Europe, particularly in France and Britain, are forcing through some of these changes. They have, after all, to make a profit for their shareholders. Privatization has not been without cost. There have been outraged stories in the British tabloids about "fat-cat water plutocrats" paying themselves multimillion-pound salaries while threatening to disconnect the urban poor who cannot pay the newly inflated rates. After privatization, the price consumers paid for water went up 106 percent, while corporate profits increased some 690 percent, and there was considerable evidence that the companies were not investing sufficiently in an aging infrastructure, with the possibility that areas of Britain would be left without enough water in a prolonged drought. But even from a social democrat point of view, the increased price of water (though not the fat-cat salaries or the profits) is the right change. It is important to signal water's value, and to price it so that waste hurts. The political trick is to balance conservation incentives against the irreducible needs of the urban poor.

"The point," says Strong, "is that water is one of those issues that no one country can solve in isolation. They have to be solved transnationally. We have to build institutions that transcend national governments."

Water survival strategy 3: *By definition, water use will go down if there are fewer people. But is this likely to happen? Was Thomas Malthus, the doomsayer who forecast worldwide famines because populations were growing faster than the food supply, right after all?*

Some years ago, in the Kenyan town of Narok, I remember chatting with a storekeeper about his family. He had ten children "so far," five with one wife and five with another.

Had the first wife died?

"No, she was used up. So I put her aside and got another."

In 1997 Kenya had 6.7 live births per woman, and the population was growing at 3.3 percent, the world's fastest, according to the Kenya National Museum in Nairobi. Without a change in the birthrate, even if the economy grows at a respectable rate, Kenya is likely to become poorer. Water poorer, too.

The museum has a permanent display on the population problem, which it calls the country's most serious — more serious by far even than AIDS. There are display cases showing the various birth control methods, and exhibits explaining the procreation process. Busloads of eager schoolchildren, in their British-legacy school uniforms, tour the exhibits every day. The museum regards this educational task as far more important than the history and wildlife displays in other exhibition rooms.

I told the curator of the population exhibit what I had found in Johannesburg. There, the black population has historically been growing at a much faster rate than the white. Popular expectations were that it would continue to do so, and that population growth would continue to worry the politicians. But, to the experts' surprise, the black birthrate has fallen dramatically in recent years, to a point where it is now just at replacement level.

"Yes," the curator said, "this is because all the girls are going to school. Teach the girls, and the problem is halfway to being solved. It can happen here too." She looked out at the chattering, eager schoolchildren. "Too bad," she said. "What they really come here to see is Ahmed." (Ahmed is the skeleton of the giant elephant that became a national institution in the 1970s; Jomo Kenyatta had even protected him through a special presidential decree.)

I looked outside where Ahmed stood, stuffed and looking desiccated, trunk reaching for the sky, tusks the color of old tea. She was wrong. The boys wanted to see Ahmed, true, but the girls were clustered around the birth control cases. Their small black faces, round and eager, were rapt. Boys will be boys, but these girls would control the future. This is accepted as a truism all over Africa and the developing world: educate the girls, and women will have fewer babies.

Population projections are a familiar source of gloomy prognostications, not only from Malthus but in recent times from people such as Paul Ehrlich, author of *The Population Bomb*. Given the startling expansion of human populations, particularly in developing countries, it seems easy to agree. Ehrlich himself has gone even further and has become an unashamed advocate of enforced birth control; and another Green doomsayer, Garrett Hardin, has been quoted as declaring that "the freedom to breed is intolerable," a pristine example of eco-fascist thinking.[13] But in 1996 United Nations population projections showed some unexpected changes. They forecast that in 2050, the world population would be smaller by nearly half a billion than earlier forecasts had shown — a decrease attributed to what is called "significant reductions in birthrates throughout the world." The population curve was flattening out — without coercion. According to figures subsequently calculated by Population Action International (PAI), an NGO whose raison d'être was slowing world population growth, this decline had some interesting ecological consequences: it meant that "there will be between 400 million and 1.5 billion fewer people living in water-short countries in the year 2050 than previously projected." This was the good news. The bad news was still bad: "Even under this improved scenario, renewable freshwater scarcity will continue to remain a problem for millions of people around the world." PAI projects that, by 2050, the percentage of the world's population living in water-stressed countries will increase by anywhere from three- to fivefold.[14] A Consultative Group on International Agricultural Research (CGIAR) report in 1998 quoted Nobel laureate Norman Borlaug as believing that the "average yields of all major food crops must increase by 50 percent by 2025 if food needs are to be met." The report also quoted the FAO as saying that two-thirds of the growth in agricultural production must come from lands already in production.[15]

Still, the lower birthrate means that certain countries — Sri Lanka or El Salvador, for instance — will delay falling into "water stress" by at least a decade, ten years in which conservation measures could be imposed. In India the change would be even more dramatic. Earlier population forecasts had shown India becoming water stressed as early as 2015. "Under the new UN projections, however, it is conceivable that India will not cross the water stress benchmark until 2035. . . . More important, if its total fertility rate were to fall in accordance with the low projection, India's population could actually begin to decline towards the middle of next century, bringing the country back out of water stress within a decade."

China is a special case, as we have seen, mostly because China is so huge. But there are some short- and medium-term solutions even there, including water diversions from the Yangtze Basin, use of less water-intensive crops, and conservation, the last oasis, as well as the adoption of local-emitter (drip) irrigation for water-intensive crops, more efficient sprinkler systems as opposed to furrow irrigation, and higher pricing for water in the cities. As China's cumbersome industrial facilities modernize, they will use less water: modern factories in the United States and Europe use less than one-third of the water for the same industrial output as Chinese factories do, and China's headlong modernization will incorporate new techniques. Chinese planners can separate industrial and household wastewater streams and use recycled household water for irrigation. They might opt for chemical or heat-treated sewage rather than employing water-borne methods. All this will buy time. But 1.5 billion relatively affluent Chinese will use a lot more water than a billion not so affluent Chinese. And that water is just not available.

Thus the Malthusians, the gloomy prognosticators of doom, are back for one more kick at the can.

Water survival strategy 4: *Steal water from others.*

In ancient Jewish, Christian, and Islamic traditions, the ultimate source of the waters of life lie beneath that politically potent piece of real estate called Jerusalem, a metaphor for the recognition that the solution to the problems of water is ultimately political. Who owns water? Who processes it? Who controls it? Who wants to steal it? Who can?

In transnational water disputes, which is the most dangerous? When the upstream nation is more powerful than the downstream, and therefore more cavalier about taking into account downstream needs? When the downstream nation is more powerful, in which case the upstream nation risks retaliation for any careless handling of the supply? Or when both countries are water stressed and more or less equal in power?

The pessimists will say all three are dangerous. Egypt, a powerful downstream riparian, has several times threatened to go to war over Nile water; only the fact that both Sudan and Ethiopia have been wracked by civil war and are too poor to develop "their" water resources has so far prevented conflict. In the Euphrates Basin, Turkey is militarily more potent than Syria, but that hasn't stopped the Syrians from threatening violence. And there are endless examples of powers that are similar in military might but have threatened war: along the Mekong River, along the Paraná, and in other places. In the Senegal Valley of West Africa, water shortages have contributed to violent skirmishes between Mauritania and Senegal, complicated by the ethnic conflict between Mauritania's black Africans and the paler-skinned Moors who control the country.

There are those who think the threats overblown. The Canadian security analyst Thomas Homer-Dixon, a name that pops up as a footnote in innumerable academic papers, is one of the skeptics. Homer-Dixon's research found virtually no examples of state violence associated with renewable resources such as fish, forests, or water, but many associated with nonrenewables such as oil or iron. He pooh-poohs the alarmists, though he acknowledges that

> water supplies are needed for all aspects of national activity, including the production and use of military power, and rich countries are as dependent on water as poor countries are. . . . Moreover, about 40 percent of the world's population lives in the 250 river basins shared by more than one country. . . . But the story is more complicated than it first appears. Wars over river water between upstream and downstream neighbors are likely only in a narrow set of circumstances. The downstream country must be highly dependent on the water for its national well-being; the upstream country must be able to restrict the river's flow; there must be a history of antagonism between the two countries; and, most important, the downstream country must be militarily much stronger than the upstream country.

He found only one case that fit all his criteria: Egypt and the Nile.

Not everyone agrees with this analysis, thinking it overly optimistic. Frederick Frey, a political scientist with the University of Pennsylvania, argues that water is different from renewable resources such as fish or wood.

> Water has four primary characteristics of political importance: extreme importance, scarcity, maldistribution, and being shared. These make internecine conflict over water more likely than similar conflicts over other resources. Moreover, tendencies towards water

conflicts are exacerbated by rampant population growth and water-wasteful economic development. A national and international "power shortage," in the sense of an inability to control these two trends, makes the problem even more alarming.[16]

There is another way of looking at the notion of water conflicts, which Homer-Dixon acknowledges and urges on the world's policymakers. Water shortages may not lead to war, but they most certainly lead to food shortages, increased poverty, and the spread of disease. They make people poorer. They increase the migrations of peoples, further straining the massive megaslums of the developing world. Standards of living deteriorate, and social unrest and violence increase, leading, as the doomsayer Robert Kaplan put it, to "the coming anarchy." Bangladesh may never go to war with India — the Bangladeshis are probably too poor to do much more than grumble — but the stress caused by water shortages has led to massive migrations of people, upsetting the ethnic balance of several Bangladeshi and Indian states, and leading to the rise of terrorist and nascent revolutionary movements.[17]

By other definitions, then — water wars.

The first question I asked when I started this book, "Is water in crisis?" was simple to pose but elusive of answer, its meanings on the one hand hidden in polemic and propaganda, and on the other rooted in the difficulties of defining any ecology (with its infinite linkages, causes shading off into consequences, sudden upwellings of pattern from apparent chaos). There are many people who seem to take an almost pornographic interest in the imminence of ecological collapse, and many, by contrast, who stubbornly refuse to look it in the snout, though it be about to bite them on the lips. I listened to the tree huggers and the apostles of free marketism (the romance of apocalypse on the one hand, a fatal narrowness of vision on the other). I listened to people comfortable with the vocabulary of crisis (Lester Brown of the WorldWatch Institute), and to those who eschew it (Peter Gleick and the ever thoughtful Sandra Postel). But in the end, for me, the central figure in the water universe was a man I came to think of as almost outside ideology, the Russian Igor Shiklomanov.

This is no amateur hydrologist. Shiklomanov has been diligently collecting water information for decades. He was beavering away long before the ruble collapsed, long before Boris Yeltsin, long before Gorbachev and perestroika, far back into the stultifying days of Soviet bureaucracy, now dismissed by the code phrase "the period of stagnation." Shiklomanov's data banks of water, of snow and ice, are used and trusted by researchers all over the globe. He is careful, meticulous, scrupulous in pointing out shortcomings in his own work. I've heard him talk a dozen times, I've read a fair proportion of his formidable output, and I have yet to catch him in hyperbole or exaggeration. He never cries doom. He never uses the word "crisis." He never speaks of precipices over which we might tumble. He hardly uses metaphors at all. He never scolds or harangues. He simply collects numbers, which he presents straightforwardly, in a genial, avuncular manner, without any embellishment. Because of this, I found him much more disturbing — even frightening — than all the many predictions of Apocalypse Soon. For what his columns of numbers and his charts and graphs show is the steadily narrowing gap between clean water supply and water demand, between water for drinking and water for food and water for sanitation and water for industry. His figures show more clearly than any polemic that water is bound up with matters of money and management and politics, and that solutions to the looming crisis in fresh water (my phrase, not his) are only to be found in the collective political will (my phrase again, definitely not his — the word "collective" has many vibrations to a Russian ear, most of them bad). In Maurice Strong's phrase, "not world government, no, but a world system."

Still, whenever my reading on water threatened to overwhelm me, either with gloom (yet another river fatally polluted, another groundwater table irretrievably and inexcusably exhausted, another million or so hectares poisoned beyond repair, another screed on the dying oceans) or with tedium (yet another academic research paper rivetingly titled something like "The Hydrological Characteristics of Schist Formations in the Permian Underlayers of the Modder River Catchment Area"), I thought of Feodor Zibold and his dew collector, and felt a little better. For what Zibold stands for, in all his fecklessness and stubbornness, is the enduring fact of human ingenuity. Out of nothing — out of folk legend and entirely erroneous guesses — Zibold created something. Water wars might be caused by human folly, but they might still be prevented by human inventiveness. What Zibold — a curious stand-in for humankind — tells us is that we are not without weapons in these wars we are waging against our own worst nature.

Notes
Bibliography
Index

Notes

1. Water in Peril

The Botswana and the "on-the-ground" passages in this chapter are taken from personal observations and interviews. The South African paper, the *Mail & Guardian,* through its Web site, has been keeping close watch on the southern African water situation, and several quotes are from that source. Material has also been drawn from an excellent study called *Water Management in Africa and the Middle East* by Kathy Eales, Simon Forster, and Lusekelo Du Mhango, quoted by Population Action International. A few statistics are drawn from Leif Ohlsson's *Hydropolitics,* an excellent compilation of global water problems by a variety of experts. And one quote is from Sandra Postel's *Last Oasis,* one of the indispensable books for anyone interested in global water issues. Selected academic papers used as sources include "Techniques for Assessing Changing Water Resources in the Twenty-first Century" by A. Rango of the USDA Hydrological Laboratory, Beltsville, Md.; and "The Consequences of Water Scarcity: Measures of Human Well-Being" by Peter Gleick. Almost all the good work on Australia seems to have been produced by John J. Pigram, director of the Centre for Water Policy Research at the University of New England, in Armidale, New South Wales. Among his papers are "Water Reform: Is There a Better Way?" and "Water Reform and Sustainability: The Australian Experience."

1. Some of these passages adapted from Marq de Villiers and Sheila Hirtle, *Into Africa* (Key Porter Books, and London: Weidenfeld & Nicolson, 1997).

2. *Mail & Guardian,* November 1996.
3. From Population Action International, largely quoting the study *Water Management in Africa and the Middle East.*
4. *Mail & Guardian,* December 1996.
5. Frank Herbert, *Dune* (New York: Ace Books, 1990), Appendix 1, "The Ecology of Dune."
6. Pigram, "Water Reform and Sustainability: The Australian Experience."
7. Figures from Ohlsson, *Hydropolitics.*
8. Postel, *Last Oasis,* p. xiv.
9. George Rothschild, director general, International Rice Research Institute, Manila, 1997.

2. The Natural Dispensation

There are numerous references and texts on the hydrological sciences. I have relied on the updated on-line *Encyclopaedia Britannica,* whose data have proved reliable, as well as the studies of Igor Shiklomanov's State Hydrological Institute in St. Petersburg, and Peter Gleick's excellent compilation, *Water in Crisis.* Academic papers include "Groundwater and Its Uses in the World" by I. S. Zektser and "Over-exploitation of Groundwater" by M. Ramon Llamas of the University of Madrid. The "second update" of Population Action International's pamphlet *Sustaining Water, Easing Scarcity,* by researchers Tom Gardner-Outlaw and Robert Engelman, was essential. Technical papers include "Water Availability and Scarcity in Central America and the Caribbean Region" by the hydrologist A. V. Izmailova, St. Petersburg; "Outlines on the Water Crisis in Latin America" by Alda da C. Rebouças, São Paulo; "The Comprehensive Assessment of the Freshwater Resources of the World: Policy Options for an Integrated Sustainable Water Future" by Johan Kuylenstierna et al.; "Global Renewable Water Resources" by Igor Shiklomanov and "World Water Resources" also by Shiklomanov, a summary of his monograph for the UN by the same name; "Principles for Assessment and Prediction of Water Use and Water Availability in the World" by Shiklomanov and N. V. Penkova; "The World Atlas of Snow and Ice" by V. M. Kotlyakov and N. M. Zverkova; "Digital Atlas of World Water Balance" by Daene C. McKinney et al., University of Texas (also available at the Web site: www.ce.utexas.edu/prof/maidment/atlas/atlas.htm); "Water Supply, Sanitation, and Environmental Sustainability: The Financing Challenge" by Ismail Serageldin, a World Bank publication;

"Water Quality for Human Needs: Criteria, Standards, and Needs" by Richard D. Robarts, Saskatoon; and "Main Water Users and Assessments of Water Withdrawals" by Mathieu Bousquet et al.

1. Larsen B study by U.S. National Ice and Snow Data Center, 1998; Larsen A study by David Vaughan and Christopher Doake of the British Antarctic Survey, recounted in *Nature* (February 1998). Some details on glacier movement from Britannica Online.
2. *Encyclopaedia Britannica,* s.v. "water."
3. Ibid.
4. Ibid.
5. *National Geographic* (March 1998).
6. *The Times,* August 28, 1999.
7. *National Geographic* (September 1999).
8. *National Post,* August 27, 1999.
9. A good discussion of the hydrosphere can be found in *Encyclopaedia Britannica.*
10. Britannica Online, "Hydrologic Cycle," http://www.eb.com:180/cgi-bin/g?DocF=micro/283/98.html.
11. Britannica Online, "The Hydrosphere: Distribution and Quantity of the Earth's Waters," http://www.eb.com:180/cgi-bin/g?DocF=macro/5002/99/0.html.
12. Adapted from Britannica Online, "The Hydrosphere: The Hydrologic Cycle," http://www.eb.com:180/cgi-bin/g?DocF=macro/5002/99/8.html.
13. From an editorial in *WorldWatch* (July/August 1983), 3.
14. Britannica Online, "Groundwater."
15. Britannica Online, "The Earth Sciences: Hydrologic Sciences: Study of the Waters Close to the Land Surface: Evaluation of the Catchment Water Basin: Groundwater."

3. Water in History

The historical anecdotes and quotes were collected from numerous sources. Particularly useful on biblical and Koranic citations was Joyce Shira Starr, whose book *Covenant over Middle Eastern Waters* contains excellent historical material. Stephen McCaffrey's survey of international water law, which appears in Peter Gleick's *Water in Crisis,* was a source for my own summary (though he is not to be blamed for conclusions I may have extracted). Roman engineering is well summarized in Trevor Hodge's book

Roman Aqueducts and Water Supply, published in 1989. Academic papers I found useful included "Water Conservation Techniques and Approaches" by D. Prinz of the University of Karlsruhe, which includes an excellent summary of ancient techniques, such as Tunisian meskats, qanats, and the collection of fog and dew. Two engaging books about attitudes to water in ancient times are *The Legendary Lore of the Holy Wells of England* by R. C. Hope, and *The Folklore of Wells,* cited below.

1. Details from John Sutton, *A Thousand Years of East Africa* (Nairobi: British Institute in Eastern Africa, 1992).
2. *Larousse Encyclopedia of Mythology* (London: Hamlyn, 1969), p. 11.
3. Camille Talkeu Tounounga, *Liquid of the Gods: Dogon Creation Myths* (UNESCO Courier, May 1993).
4. Sir Rustom Pestonji Masani, *Folklore of Wells* (Norwood, Pa.: Norwood Editions, 1974).
5. *Encyclopaedia Britannica,* s.v. "groundwater."
6. Stephen McCaffrey, "Water, Politics, and International Law," in Gleick, *Water in Crisis,* p. 99.
7. *The Economist,* July 19, 1997.
8. *National Geographic* (November 1993), 30.

4. Climate, Weather, and Water

Passages on Timbuktu, the Namib Desert, and other places were drawn from personal observations. The post-Kyoto proceedings of the Eminent Persons Group on Climate Change were useful, as was the ongoing debate in both technical and popular journals about CO_2 production and its effects on climate. The Israeli Government Information Office is a cornucopia of interesting material on combating desertification. One useful paper was Maria C. Donoso's "Assessment of Climate Variability Impact on Water Resources in the Humid Tropics of the Americas." Other books and papers include "Country-Specific Market Impacts of Climate Change" by Robert Mendelsohn et al. of Yale University; *Meeting the Challenge of Population, Environment, and Resources: The Costs of Inaction* by Henry Kendall et al. of the Union of Concerned Scientists; "Global Climate Warming Effects on Water Resources" by V. Yu. Georgiyevsky; "Induced Climate Change Impacts on Water Resources" by Eugene Stakhiv of the U.S. Army Corps of Engineers; and "Water, Climate Change, and Health" by Paul R. Epstein of Harvard Medical School.

1. Reed P. Shearer et al., in *Science* (July 1998).
2. For a discussion of the Sahara, see Tor Eigeland et al., *The Desert Realm: Lands of Majesty and Mystery* (Washington, D.C.: National Geographic Society, 1982), p. 112.
3. Britannica Online, "The Biosphere and Concepts of Ecology: Major Ecosystems of the World: Terrestrial Ecosystems: Deserts: Environment," http://www.eb.com:180/cgi-in/g?DocF=macro/5000/74/155.html.
4. *National Geographic* (November 1979).
5. Israeli Government Information Office fact sheet (untitled). For some of Israel's recommended solutions, see Chapter 16, "Solutions and Manifestos."
6. Magnus Magnusson and Hermann Pálsson, *The Vineland Sagas: The Norse Discovery of America* (London: Penguin, 1965).
7. *The Economist Survey of Development and the Environment,* March 21, 1998.
8. Reported in *The Economist.*
9. Data from "Water, Climate Change, and Health," a paper by Paul R. Epstein, associate director of the Center for Health and the Global Environment at Harvard Medical School.
10. Reported by Traci Watson, *USA Today,* October 29, 1998.
11. *The Economist,* November 7, 1998.
12. Eugene Stakhiv, "Induced Climate Change Impacts on Water Resources" (U.S. Army Corps of Engineers, June 1998).
13. International Development Research Center Home Page.

5. Unnatural Selection

The sources for this chapter are too numerous to list all of them. Some data for the Volga River, eastern Europe, the U.S.-Canadian Great Lakes, and other places is from personal observations and interviews. Particularly useful were Linda Nash's survey, "Water Quality and Health," in Peter Gleick's *Water in Crisis,* and reports from the NGO WaterAid, particularly a thorough report by researcher Maggie Black. Other sources include a case study called "Water Supply of Middle-sized Latin American Cities Endangered by Uncontrolled Groundwater Exploitation in Urban Areas" by J. Bundschuh et al. of Salta University in Buenos Aires. There are excellent "pollution maps" in the *National Geographic's* revised sixth edition of the *Atlas of the World* (1995), from which some of the pollution data were drawn.

1. M. Meybeck, "Surface Water Quality: Global Assessment and Perspective" (1998).
2. *The Times,* October 26, 1998.
3. National Wildlife Federation Fact Sheet, undated.
4. Details from Linda Nash, "Water Quality and Health," in Gleick, *Water in Crisis,* p. 25.
5. Report commissioned by *USA Today,* October 1998.
6. Quoted in Nash, "Water Quality and Health," p. 31.
7. F. B. Smith, *The People's Health, 1830–1910* (London: Croom Helm, 1979), quoted in Maggie Black, *Mega-Slums* (WaterAid, 1998).
8. Details of this discussion of megacities come from many sources, but especially Maggie Black's excellent WaterAid report, *Mega-Slums.*
9. Ibid.
10. Nash, "Water Quality and Health," p. 27.
11. *Britannica Online,* "The Earth Sciences: Hydrologic Sciences: Study of the Waters Close to the Land Surface: Water Quality," http://www.eb.com:180/cgi-bin/g?DocF=macro/5001/89/92.html.

6. The Aral Sea

Descriptions of the region are from personal observations and visits. The best journalistic account I have seen was by Don Hinrichsen in *People & the Planet* magazine, cited below. Also useful was Arun P. Elhance's "Conflict and Cooperation over Water in the Aral Sea Basin," in *Studies in Conflict and Terrorism,* 20 (1997), 207–218.

1. The details of this action program were presented to a UNESCO conference in Paris in June 1998. The paper was titled "The Future of the Aral Sea Basin: How Can Independent States Survive in Water Scarcity?"
2. David R. Smith, "Environmental Security and Shared Water Resources in Post-Soviet Central Asia," in *Post-Soviet Geography* (June 1995).
3. Reported by Don Hinrichsen in *People & the Planet* 4, no. 2 (1995).

7. To Give a Dam

Marc Reisner's *Cadillac Desert,* which is almost compulsively readable, contains much fascinating data, especially about U.S. dams, but also about the siltation problem faced by all dams. Janet Abramovitz's 1996 study *Imperiled Waters, Impoverished Future* is thorough and thoughtful, understated but chilling. The Internet resounds with polemics on dams. Especially useful is the RiverNet Web site. A good summary of the Three Gorges project from a Chinese perspective is C. Yangbo's paper "The Three Gorges Project: Key Project for Transferring Water from South to North" (Yichang, China, 1998).

1. Reisner, *Cadillac Desert,* p. 469.
2. Peter Theroux, "The Imperiled Nile Delta," *National Geographic* (January 1997).
3. Janet Abramovitz, *Imperiled Waters, Impoverished Future* (World-Watch, 1996).
4. "Athabasca Chipewyan First Nation Inquiry: W. A. C. Bennett Dam and Damage to Indian Reserve 201 Claim," in *Indian Claims Commission Proceedings* 10 (1998), 117ff.
5. Reisner, *Cadillac Desert.*
6. Abramovitz, *Imperiled Waters, Impoverished Future,* pp. 15ff.
7. Philip Fradkin, *A River No More,* pp. xiv and 199.

8. The Problem with Irrigation

Sandra Postel has said many sensible things about irrigation in various publications, especially her book *Last Oasis* and her survey "Water and Agriculture," in Peter Gleick's *Water in Crisis.*

1. Much of the data in this section comes from Sandra Postel's excellent survey "Water and Agriculture," in Gleick, *Water in Crisis.*
2. Eigeland, *The Desert Realm,* p. 65.
3. Figures from the U.S. Salinity Laboratory, Riverside, California.
4. Postel, "Water and Agriculture," p. 58.

9. Shrinking Aquifers

Notes and observations from personal visits to North Africa and the American Southwest form the basis for this chapter. The Libyans have produced a number of interesting studies of the Great Man-Made River Project, among them a useful technical paper, "Man-made Rivers: A New Approach to Water Resources Development in Dry Areas," by the hydrologist Saad al-Ghariani, professor of water sciences at Al Fateh University in Tripoli. Much useful material was gleaned from technical papers such as "The Aquifers of the Grand Basins: A Vital Resource for Development and for the Struggle against Desertification in the Arid and Semi-arid Zones" by Jean-Marc Louvet of the Observatoire du Sahara et du Sahel in Paris; "Socioeconomic Aspects of the Demand for Domestic Water in Sub-Saharan Africa: What Lessons for Water Management?" by Étienne J. Maïga et al.; and "The Dwindling Water Resources of Lake Chad: An Alarming Trend of the Twenty-first Century" by Abubakar B. Jauro of the Lake Chad Basin Commission, Ndjamena. Marc Reisner, again, was an indispensable source on the Ogallala Aquifer. So was the report edited by Michael Glantz, *Forecasting by Analogy: Societal Responses to Regional Climatic Change* (1989), whose summary on the Ogallala was based on a study by Donald A. Wilhite of the Center for Agricultural Meteorology and Climatology, University of Nebraska at Lincoln.

1. Britannica Online, "Sahara: Physical Features," http://www.eb.com: 180/cgi-n/g?DocF=macro/5000/3/81.html.
2. Eigeland, *The Desert Realm,* p. 120.
3. Quoted in James H. Gray, *Man against the Desert* (Saskatoon: Western Producer Book Service, 1967).
4. Ibid.
5. For a much more detailed and highly readable account of the Ogallala problem, see Reisner, *Cadillac Desert.*

10. The Reengineered River

The recovery of the Rhine has been widely reported. A useful summary is contained in a report for the International Wildlife Federation by Page Chichester (cited below), which was supplemented by my own visits and interviews. The International Commission for the Protection of the Rhine has its own Web page, which is surprisingly candid. A useful summary paper is Aleta Brown's "Bringing Rivers Back from the Brink," for the International Rivers Network. The Danube story was put together from personal observations and from the proceedings of the International Court of Justice at The Hague. Two papers by the Russian hydrologist I. P. Zaretskaya were particularly helpful: "Water Availability and Use in the Danube Basin" and "State of the Art: Expected Water Availability and Water Use in the Danube Basin." The Hungarian Greens have been very active, and their discussions on the Internet proved very useful.

1. Quoted in Page Chichester, *International Wildlife* (National Wildlife Federation, 1997).
2. Ibid.
3. Chris Bright, *WorldWatch* (May/June 1999), p. 12.

11. The Middle East

Much of this chapter was produced after on-the-ground observations and interviews with many of the participants. Endless amounts of material are available on Middle East water policies and politics, both in print form and on-line. The various water plans cited are all easily available from many sources. Helena Lindholm's study (cited below) in Leif Ohlsson's *Hydropolitics* provides an intelligent political overview. Other studies include "The Water Sources of the West Bank of the Jordan Valley and Their Utilization" by Aharon Yaffe, Jerusalem; "Water Resources Scarcity in Yemen: A Time Bomb under Socioeconomic Development" by Jac A. M. van der Gun; "A Sober Approach to the Water Crisis in the Middle East" by Jad Isaac for the Applied Research Institute, Jerusalem; "Troubled Water of Eden" by Daniel Hillel for People and the Planet (a reworking of his excellent book *The Rivers of Eden: The Struggle for Water and the Quest for Peace in the Middle East,* published by Oxford University Press in 1994); "Roots of the Water Conflict in the Middle East" by Jad Isaac and Leonardo Hosh, delivered at the University of Waterloo, Ontario, in 1992; and *The Politics of Scarcity: Water in the Middle East* by Joyce Starr and Daniel Stoll (Westview Special Studies series on the Middle East, 1988).

1. Thomas Naff, quoted by Helena Lindholm in "Water and the Arab-Israeli Conflict," in Ohlsson, *Hydropolitics*, p. 76.
2. Quoted by McCaffrey, "Water, Politics, and International Law," p. 92.
3. Nir Becker and Naomi Zeitouni, "A Market Solution for the Israeli-Palestinian Water Dispute," *Water International* (December 1998).

12. The Tigris-Euphrates System

James Cran (see Chapter 16) was useful on Turkish water politics. Turkish academic studies largely agree with Syrian concerns about what the GAP project will mean to Syrian water supply, but their authors are not always willing to go public with their beliefs, since the GAP has an irresistible political momentum in Turkey.

1. Quoted by Amikam Nachmani in "Water Jitters in the Middle East," *Studies in Conflict and Terrorism* 20.1 (1997), 67.
2. Figures from Turkey's State Hydraulic Works show per capita water availability at 2,110 cubic meters per year for Iraq, 1,830 for Turkey, 1,420 for Syria, 300 for Israel, 250 for Jordan, and 100 for Palestine.
3. Ilter Turan, *Turkey and the Middle East: Problems and Solutions* (Water International, 1993).
4. Mikhail Wakil, "Analysis of Future Water Needs for Different Sectors in Syria," *Water International* 18.1 (1993).
5. Quoted by McCaffrey, "Water, Politics, and International Law," p. 93.

13. The Nile

I traveled virtually every kilometer of the Nile, and the descriptions come from those personal visits to Egypt, Sudan, Ethiopia, Uganda, and Kenya. Jan Hultin's study "The Nile: Source of Life, Source of Conflict," in Leif Ohlsson's *Hydropolitics,* is a good start for anyone interested in the subject. Many of the ancient legends cited were collected by the indefatigable if atrocious Henry Morton Stanley, who at least on these has proved reliable. Other sources include "Water Resources Assessment in South Sinai, Egypt" by A. M. Amer et al. and "National Environment Measures and Their Impact on Drainage Water Quality in Egypt" by M. A. Abdel-Khalik et al., Cairo.

1. Henry Morton Stanley, *In Darkest Africa* (London: Samson Kow, Marston, Searle and Rivington, 1890), p. 448.
2. Ibid.
3. Postel, *Last Oasis,* citing a study by Dale Whittington and Elizabeth McClelland of the University of North Carolina.

14. The United States and Its Neighbors

Again, Marc Reisner's *Cadillac Desert* is must reading for anyone interested in American water politics. Sandra Postel, in *Last Oasis,* and in her PBS miniseries, is particularly good on the lower Colorado River. Other sources include *Water Markets: Priming the Invisible Pump,* written by Terry Anderson and Pamela Snyder for the Cato Institute; *A Life of Its Own: The Politics and Power of Water* by Robert Gottlieb, excellent on the California water cartels; a *Wall Street Journal* piece by Mitchel Benson on water law (October 1, 1997); and various articles from U.S. Water News Online. The Canadian Environmental Law Association, based in Toronto, is an indispensable source for anyone interested in Canadian water politics, and the way the Canadian authorities are fumbling the transition to privately run water utilities and the handling of water under the NAFTA rules.

1. Reisner, *Cadillac Desert,* p. 484.
2. Quoted by Postel, *Last Oasis,* p. ix.
3. U.S. Water News Online, June 1977.
4. Clark S. Knowlton, "International Water Law along the Mexican-American Border," paper given to a symposium of the American Association for the Advancement of Science, El Paso, April 1968.
5. Reisner, *Cadillac Desert,* p. 7.
6. *Los Angeles Times,* December 18, 1998.
7. *The Economist,* January 24, 1998.
8. *Toronto Globe and Mail,* May 23, 1998.
9. J. C. Day, Kristan M. Boudreau, and Nancy C. Hackett, "Emerging Institutions for the Bilateral Management of the Columbia River Basin," *American Review of Canadian Studies* (Summer 1996).
10. *The Economist,* March 29, 1997.

15. The Chinese Dilemma

Coverage of China's water woes was given a huge impetus in the summer of 1998, when the Yangtze catastrophically overflowed, leading to endless charges and countercharges about the role in the disaster played by the Three Gorges Dam project, under construction. For the first time, that summer saw the emergence of an activist Green movement inside China. Much of the rest of the debate, about China's food production, was set off by a paper by the veteran environmentalist Lester Brown, *Who Will Feed China?* which was indignantly attacked within China.

1. Statistics from Lester Brown, *Who Will Feed China?*
2. Ed Ayres in *WorldWatch* (November/December 1998).
3. CGIAR was established in 1971 as a quasi-NGO under UN auspices "to support productivity-oriented research in response to food needs of near-famine proportions in the South."

16. Solutions and Manifestos

The water world seems to be split between those who believe that there are technical (that is, engineering and scientific) solutions to the looming crisis in fresh water and those who believe that the problem is more one of politics and management. Sandra Postel leans to the latter, as does Maurice Strong, who was secretary general of the Rio conference on the environment in the early 1990s, and who has been active in water politics ever since, partly through the Earth Council, which he founded, and through his influence on the World Bank. Of course, the two positions overlap, as I have tried to show: the best conservation methods come from finely tuned engineering. It's why people like Maurice Strong have argued so vigorously for the international sharing of scientific data. Other sources include papers such as "Transboundary Fresh Water Disputes and Conflict Resolution: Planning an Integrated Approach" by Edy Kaufman et al., in *Water International;* "The Economic Value of Water" by Kindler Janusz, Warsaw; and "Water and Food" by W. E. Klohn et al. of FAO (this includes useful sections on, for example, how much water it takes to produce food).

1. Wade Graham, "A Hundred Rivers Run Through It," *Harper's* (June 1998).
2. Nachmani, "Water Jitters in the Middle East," pp. 67–93.
3. *The Economist,* August 21, 1999.
4. Details from the California Water Commission.
5. Woodward-Clyde Consultants, environmental impact report for the City of Santa Barbara and Ionics, Inc.'s, Temporary Emergency Desalination Project, March 1991.
6. Quoted by David Lees, "Food by Design," *Financial Post* (October 1998).
7. *The Economist,* August 21, 1999.
8. Postel, *Last Oasis,* p. 129.
9. Ibid., p. 129, citing Shaul Streit, "On-land Treatment and Disposal of Municipal Sewage: Agro-Sanitary Integration," paper for a World Bank seminar, 1992. Postel's book contains a good survey of wastewater treatment.
10. Postel, *Last Oasis,* p. 135.
11. Sheldon Rampton in *Harper's* (November 1998).
12. André de Moor and Peter Calamai, "Subsidizing Unsustainable Devel-

opment: Undermining the Earth with Public Funds." Earth Council, 1997.

13. *The Economist,* December 20, 1997.

14. Tom Gardner-Outlaw and Robert Engelman, *Sustaining Water, Easing Scarcity: A Second Update,* Population Action International, Washington, D.C., 1997.

15. Third System Review of the Consultative Group on International Agricultural Research, October 1998.

16. Frederick W. Frey, "The Political Context of Conflict and Cooperation over International River Basins," *Water International* 18 (1993).

17. See particularly Thomas Homer-Dixon, *Environmental Scarcity and Global Security* (New York: Foreign Policy Association Headline Series, 1993).

Bibliography

Some of the essential sources were mentioned in my introduction. But of course there are many, many others.

Books and Pamphlets

Abramovitz, Janet N. *Imperiled Waters, Impoverished Future: The Decline of Freshwater Ecosystems.* Washington, D.C.: WorldWatch Institute, 1996.

Anderson, Terry L., and Pamela Snyder. *Water Markets: Priming the Invisible Pump.* Washington, D.C.: Cato Institute, 1979.

Basson, M. S. *Overview of Water Resources Availability and Utilisation in South Africa.* Pretoria: Department of Water Affairs and Forestry, 1997.

Brown, Lester R. *Who Will Feed China? Wake-up Call for a Small Planet.* New York: W. W. Norton, 1995.

CGIAR System Review Secretariat. *The International Research Partnership for Food Security and Sustainable Agriculture, Third System Review.* October 8, 1998.

Clarke, Robin. *Water: The International Crisis.* Boston: MIT Press, 1993.

Cossi, Olga. *Water Wars: The Fight to Control and Conserve Nature's Most Precious Resource.* New York: New Discovery Books, 1993.

Dixon Hardy, P. *The Holy Wells of Ireland.* Dublin: Hardy and Walker, 1840.

Eigeland, Tor, et al. *The Desert Realm: Lands of Majesty and Mystery.* Washington, D.C.: National Geographic Society, 1982.

Encyclopaedia Britannica (both print and on-line editions).

Engelman, Robert. *Why Population Matters.* Washington, D.C.: Population Action International, 1997.

Engelman, Robert, and Pamela LeRoy. *Sustaining Water: Population and the Future of Renewable Water Supplies.* Washington, D.C.: Population Action International, 1993.

Farid, Claire, John Jackson, and Karen Clark. *The Fate of the Great Lakes: Sustaining or Draining the Sweetwater Seas?* Buffalo, N.Y.: Canadian Environmental Law Association and United Buffalo State College, 1997.

Fradkin, Philip L. *A River No More: The Colorado River and the West.* Revised edition. Berkeley: University of California Press, 1996.

Gardner-Outlaw, Tom, and Robert Engelman. *Sustaining Water, Easing Scarcity: A Second Update.* Washington, D.C.: Population Action International, 1997.

Glantz, Michael. *Forecasting by Analogy: Societal Response to Regional Climatic Change.* 1989. (Includes a summary on the Ogallala, based on a study by Donald A. Wilhite, Center for Agricultural Meteorology and Climatology, University of Nebraska at Lincoln.)

Gleick, Peter H. *The World's Water: The Biennial Report on Fresh Water Resources, 1998–1999.* Washington, D.C.: Island Press, 1998.

———, ed. *Water in Crisis: A Guide to the World's Fresh Water Resources.* New York: Oxford University Press, 1993.

Gottlieb, Robert. *A Life of Its Own: The Politics and Power of Water.* New York: Harcourt Brace Jovanovich, 1988.

Graves, Robert, ed. *New Larousse Encyclopedia of Mythology.* London: Paul Hamlyn, 1969.

Gray, James H. *Men Against the Desert.* Saskatoon, Sask.: Western Producer Book Service, 1967.

Herbert, Frank. *Dune.* New York: Ace Books, 1990.

Hillel, Daniel. *The Rivers of Eden: The Struggle for Water and the Quest for Peace in the Middle East.* New York: Oxford University Press, 1994.

Homer-Dixon, Thomas F. *Environmental Scarcity and Global Security.* New York: Foreign Policy Association Headline Series, 1993.

Hope, R. C. *The Legendary Lore of the Holy Wells of England.* London: Elliot Stock, 1893.

Kendall, Henry, et al. *Bioengineering of Crops: Report of the World Bank Panel on Transgenic Crops.* Washington, D.C.: World Bank, 1997.

———. *Meeting the Challenges of Population, Environment, and Resources:*

The Cost of Inaction. A Report of the Senior Scientists' Panel, Union of Concerned Scientists. Washington, D.C.: World Bank, 1995.

Loude, Jean-Yves, and Viviane Lièvre. *Le roi d'Afrique et la reine mer.* Quoted in *Balafon* 126 (1996), the in-flight magazine of Air Afrique.

Lowi, Miriam R. *The Politics of Water: The Jordan River and the Riparian States.* Montreal: McGill Studies in International Development, no. 35, 1984.

Magnusson, Magnus, and Hermann Pálsson. *The Vineland Sagas: The Norse Discovery of America.* London: Penguin, 1965.

Masani, Sir Rustom Pestonji. *The Folklore of Wells.* Norwood, Penn.: Norwood Editions, 1974.

National Geographic Society. *Water: The Power, Promise, and Turmoil of North America's Fresh Water.* Special Issue, vol. 184, no. 5A. Washington, D.C., 1993.

Ohlsson, Leif, ed. *Hydropolitics: Conflicts over Water as a Development Constraint.* London: Zed Books, 1995.

Postel, Sandra. *Dividing the Waters: Food Security, Ecosystem Health, and the New Politics of Scarcity.* Washington, D.C.: WorldWatch Institute, 1996.

——. *Last Oasis: Facing Water Scarcity.* New York: W. W. Norton, 1997.

Quiller-Couch, M. and L. *Ancient Holy Wells of Cornwall.* London: Chas. J. Clark, 1894.

Reisner, Marc. *Cadillac Desert: The American West and Its Disappearing Water.* Revised and updated. New York: Penguin, 1986.

Starr, Joyce Shira. *Covenant over Middle Eastern Waters: Key to World Survival.* New York: Henry Holt, 1995.

Sutton, John. *A Thousand Years of East Africa.* Nairobi: British Institute in Eastern Africa, 1992.

Walton, K. *The Arid Zones.* London: Hutchinson, 1969.

Worster, Donald. *Rivers of Empire: Water, Aridity, and the Growth of the American West.* New York: Oxford University Press, 1985.

Young, Gordon J., James C. I. Dooge, and John C. Rodda. *Global Water Resource Issues.* Cambridge: Cambridge University Press, 1994.

Academic and Technical Papers

Papers unsourced in this list are from the Proceedings of the International Conference on Water ("Water: The Looming Crisis"), Paris, June 1998, published by UNESCO.

Abdel-Khalik, M. A., et al. (Cairo). "National Environment Measures and Their Impact on Drainage Water Quality in Egypt."

Al-Ghariani, Saad. "Man-made Rivers: A New Approach to Water Resources Development in Dry Areas." *Water International* 22.2 (1997).

Amery, Hussein A. "Water Security as a Factor in Arab-Israeli Wars and Emerging Peace." *Studies in Conflict and Terrorism* 20.1 (1997).

Bedford, D. P. "International Water Management in the Aral Sea Basin." *Water International* 21.2 (1996).

Boronkay, Carl. "Water Conflicts in the Western United States." *Studies in Conflict and Terrorism* 20.2 (1997).

Bousquet, Mathieu, et al. "Main Water Users and Assessments of Water Withdrawals."

Bundschuh, J., et al. (Salta University, Buenos Aires). "Water Supply of Middle-Sized Latin American Cities Endangered by Uncontrolled Groundwater Exploitation in Urban Areas."

Chandra, Sekhar M. (Warangal, India). "Integrated Water Quality Management Plan for Bhadri Lake, India."

de Moor, André, and Peter Calamai. "Subsidizing Unsustainable Development: Undermining the Earth with Public Funds." Chapter 2: "Water, Water Everywhere (but Not a Drop to Waste)." Earth Council, 1997.

Donoso, Maria C. "Assessment of Climate Variability Impact on Water Resources in the Humid Tropics of the Americas."

Dukhovny, Viktor (Scientific Information Committee of the Five-State Interstate Water Commission). "The Future of the Aral Sea Basin: How Can Independent States Survive in Water Scarcity?"

Eales, Kathy, Simon Forster, and Lusekelo Du Mhango. *Water Management in Africa and the Middle East.* Edited by Egial Rached et al. Ottawa: IDRC, 1996.

Elhance, Anin P. "Conflict and Cooperation over Water in the Aral Sea Basin." *Studies in Conflict and Terrorism* 20 (1997).

Epstein, Paul R. (Harvard University Medical School). "Water, Climate Change, and Health."

Falkenmark, Malin. "Global Water Issues Confronting Humanity." *Journal of Peace Research* 27.2.

———. "The Massive Water Scarcity Now Threatening Africa: Why Isn't It Being Addressed?" *Ambio* 18.2.

———. "Middle East Hydropolitics: Water Scarcity and Conflicts in the Middle East." *Ambio* 18.6.

——— "Vulnerability Generated by Water Scarcity." *Ambio* 18.6.

Frey, Frederick W. "The Political Context of Conflict and Cooperation over International River Basins." *Water International* 18 (1993).

Georgiyevsky, V. Yu. "Global Climate Warming Effects on Water Resources."

Gleick, Peter. "Basic Water Requirements for Human Activities: Meeting Basic Needs." *Water International* 21.2 (1996).

———. "The Consequences of Water Scarcity: Measures of Human Well-Being."

Güner, Serdar. "The Turkish-Syrian War of Attrition: The Water Dispute." *Studies in Conflict and Terrorism* 20.1 (1997).

Hady Rady, Mohammed Abdel. "Satisfying National and International Water Demands." *Water International* 20.1 (1995).

Hamdy, Atef, et al. "Water Crisis in the Mediterranean: Agricultural Water Demand Management." *Water International* 20.4 (1995).

Hasan, Arif. "Innovative Sewerage in a Karachi Squatter Settlement." Orangi Pilot Project, 1986.

Hillel, Daniel. "Troubled Water of Eden, for People and the Planet." A reworking of his book, *The Rivers of Eden: The Struggle for Water and the Quest for Peace in the Middle East.* New York: Oxford University Press, 1994.

Isaac, Jad (Applied Research Institute, Jerusalem). "A Sober Approach to the Water Crisis in the Middle East."

Isaac, Jad, and Leonardo Hosh. "Roots of the Water Conflict in the Middle East." Paper delivered at the University of Waterloo, Ontario, Canada, 1992.

Izmailova, A. V. (St. Petersburg). "Water Availability and Scarcity in Central America and the Caribbean Region."

Janusz, Kindler (Warsaw). "The Economic Value of Water."

Jauro, Abubakar B. "The Dwindling Water Resources of Lake Chad: An Alarming Trend of the Twenty-first Century." Lake Chad Basin Commission, Ndjamena.

Klohn, W. E., et al. "Water and Food." Study for the FAO.

Kotlyakov, V. M., and N. M. Zverkova. "The World Atlas of Snow and Ice."

Kuylenstierna, Johan, et al. "The Comprehensive Assessment of the Freshwater Resources of the World: Policy Options for an Integrated Sustainable Water Future." *Water International* 23.1 (1998).

Llamas, M. Ramon (University of Madrid). "Over-exploitation of Groundwater."

Louvet, Jean-Marc. "The Aquifers of the Grand Basins: A Vital Resource for Development and for the Struggle against Desertification in the Arid and Semi-arid Zones." Study for Observatoire du Sahara et du Sahel, Paris.

Maïga, Étienne J., et al. "Socioeconomic Aspects of the Demand for Domestic Water in Sub-Saharan Africa: What Lessons for Water Management?"

McKinney, Daene C., et al. (University of Texas). "Digital Atlas of World Water Balance." Also available at the Web site: www.ce.utexas.edu/prof/maidment/atlas/ atlas.htm.

Mendelsohn, Robert, et al. (Yale University). "Country-Specific Market Impacts of Climate Change."

Meybeck, M. (Marie M. Curie University, Paris). "Surface Water Quality: Global Assessment and Perspective."

Moraru-de Loë, Liana, et al. "Public-Private Partnerships: Water and Wastewater Services in France." *Water International* 18.3 (1993).

Morris, Mary E. "Water and Conflict in the Middle East: Threats and Opportunities." *Studies in Conflict and Terrorism* 20.1 (1997).

Murray, Dianne. "Dams and Dying Fisheries." Hudson Bay Research Centre, Carleton University, 1998.

Nachmani, Amikam. "Water Jitters in the Middle East." *Studies in Conflict and Terrorism* 20.1 (1997).

Naff, Thomas. "Information, Water, and Conflict: Exploring the Linkages in the Middle East." *Water International* 22.1 (1997).

Neu, H. J. A. "Man-Made Storage of Water Resources: A Liability to the Ocean Environment?" Part 1, *Marine Pollution Bulletin* 13.1: 7–12; part 2, *Marine Pollution Bulletin* 13.2: 44–47.

———. "Runoff Regulation for Hydropower and Its Effects on the Ocean Environment." *Canadian Journal of Civil Engineering* 2: 583–591.

Niemczynowicz, Janusz. "Megacities from a Water Perspective." *Water International* 21.4 (1996).

Pigram, John J. (Centre for Water Policy Research at the University of New England, Armidale, N.S.W., Australia). "Water Reform: Is There a Better Way?"

———. "Water Reform and Sustainability: The Australian Experience." Paper for the Centre for Water Policy Research, Ninth World Water Congress, Montreal, September 1997.

Postel, Sandra. "Running Dry: Escalating Consumption Is Threatening to Outpace Water Supply." *UNESCO Courier,* May 1993.

Prinz, D. (University of Karlsruhe). "Water Conservation Techniques and Approaches." Includes an excellent summary of ancient techniques, such as Tunisian meskats, clanats, and the collection of fog and dew.

Rango, A. (USDA Hydrological Laboratory, Beltsville, Md). "Techniques for Assessing Changing Water Resources in the Twenty-first Century."

Rebouças, Alda da C. (São Paulo). "Outlines on the Water Crisis in Latin America."

Robarts, Richard D. (Saskatoon, Sask., Canada). "Water Quality for Human Needs: Criteria, Standards, and Needs."

Roberts, Bruce R. "Water Management in Desert Environments: A Comparative Analysis." *Lecture Notes in Earth Sciences* 48 (1993).

Robinson, Nicholas, ed. "Agenda 21: Earth's Action Plan" (annotated). The Commission on Environmental Law of IUCN. New York: World Conservation Union, Oceana Publications Inc., 1993.

Salmi, Ralph H. "Water, the Red Line: The Interdependence of Palestinian and Israeli Water Resources." *Studies in Conflict and Terrorism* 20.1 (1997).

Samaddar, Ranabir. "Flowing Waters and the Nationalist Metaphors: The Dispute between India and Bangladesh over the Ganges." *Studies in Conflict and Terrorism* 20.2 (1997).

Serageldin, Ismail. "Water Supply, Sanitation, and Environmental Sustainability: The Financing Challenge." Washington, D.C.: Directions in Development, World Bank, 1997.

Shady, Ali, et al. "The Nile 2002: The Vision toward Cooperation in the Nile Basin." *Water International* 19.2 (1994).

Shiklomanov, Igor. "Climate of Uncertainty: Climate Changes Resulting from Human Activity Will Have Serious Implications for the Distribution of the World's Water." *Water International* 18.5 (1993).

———. "Global Renewable Water Resources."

———. "World Water Resources." Summary of his monograph for the UN by the same name.

Shiklomanov, Igor, and N. V. Penkova. "Principles for Assessment and Prediction of Water Use and Water Availability in the World."

Singh, Udai P., and Otto J. Helweg, eds. "Supplying Water and Saving the Environment for Six Billion People." Proceedings of Selected Sessions from the 1990 ASCE Convention, 1990.

Singhal, D. C. (Roorkee, India). "Environmental Impact of Groundwater Utilization and Development: The Indian Perspective."

Smith, David R. "Environmental Security and Shared Water Resources in Post-Soviet Central Asia." *Post-Soviet Geography* (June 1995).

Staff, Joyce, and Daniel Stoll. "The Politics of Scarcity: Water in the Middle East." Westview Special Studies series on the Middle East, 1988.

Stakhiv, Eugene (U.S. Army Corps of Engineers). "Induced Climate Change Impacts on Water Resources."

Suresh, S. (Bangalore). "Intersectoral Competition for Land and Water between Users and Uses in Tamil Nadu."

Turan, Iter. "Turkey and the Middle East: Problems and Solutions." *Water International* 18.1 (1993).

Union of Concerned Scientists. "World Scientists' Call for Action at the Kyoto Climate Summit." Cambridge, Mass., 1997.

Ünver, Olcay, et al. "The Southeastern Anatolia Project — GAP: Improvement of Canal Regulation Techniques; Improvement of Field Water Distribution and Irrigation Techniques." *Water International* 18.3 (1993).

van der Gun, Jac A. M. "Water Resources Scarcity in Yemen: A Time Bomb under Socioeconomic Development."

Water Technology Board Tenth Anniversary Symposium. "Sustaining Our Water Resources." Washington, D.C.: National Academy Press, 1993.

Yaffe, Aharon (Jerusalem). "The Water Sources of the West Bank of the Jordan Valley and Their Utilization."

Yangbo, C. "The Three Gorges Project: Key Project for Transferring Water from South to North." Yichang, China, 1998.

Zaretskaya, I. P. "State of the Art: Expected Water Availability and Water Use in the Danube Basin."

——. "Water Availability and Use in the Danube Basin."

Zektser, L. S. "Groundwater and Its Uses in the World."

Index

Abramovitz, Janet, 128
Abu Simbel, 124
Accra, 52
Acid rain, 29
Adam Bridge, 190
Addis Ababa, 227
Afghanistan, 55
Africa, 60, 89, 90, 95, 97, 102, 120, 121,
    131, 132, 133, 138, 139, 163, 217, 307, 310
    pollution of, 89
African Development Bank, 225
Agadez, 146
Agarwal, Anil, 128
Ahaggar Mountains, Sahara, 70, 148,
    154
Alaska, 15, 28, 254, 256, 278, 281, 285
Aleppo, 210, 211
Alexander the Great, 185
Alexandria, 216
Al Fateh University, Libya, 150
Algae, 88, 90, 104, 170, 178, 298
    blooms, 25, 80, 170
Algeria, 14, 148, 153, 154
al-Ghariani, Saad, 150, 151, 153
Allenby Bridge (Jordan River), 196
Alsace, 88
Amazon, 38, 88, 95
Amazon Basin, 22, 37
American Journal of Epidemiology, 97
American Society of Civil Engineers,
    165

Amman, 100, 197, 198
Amu Darya River, 105, 106, 108, 109,
    110, 112
Amur, 38
Anatolian Highlands, 204, 205, 209
Anersen, Per Pinstrup, 20
Angola, 4, 5, 7, 10
Animas La Plata Dam, 303
Ankara, 208, 213
Antarctic, 28, 67, 79
Antarctica, 285
Aqueducts, 7, 12, 41, 46, 55, 56, 57, 138,
    242, 259, 291
Aquicludes, 42
Aquifers, 7, 9, 12, 14, 15–16, 21, 34, 36–
    37, 40–45, 54–55, 60, 63, 103, 140,
    148–54, 157–62, 164–65, 179, 189–91,
    196, 198, 200–201, 219, 234, 240, 247,
    253, 261, 280, 294, 299–300, 302–3
    depletion (mining of), 151ff
    overdrafting, 161ff
Arabian Peninsula, 14, 23, 164, 185
Arab-Israeli war, 190ff
Arab League, 194
Aral Basin, 106, 112, 113
Aral Sea, 105–6, 108–15, 113, 143, 159,
    163, 170, 241, 251, 255, 281
    and climate change, 110
    declared a disaster area, 112
    salvation plan for, 114
Arctic ice, 29, 81, 280–81